PERFORMANCE OF THERMAL UTILITIES IN INDIAN ENERGY SECTORS

PERFORMANCE OF THERMAL UTILITIES IN INDIAN ENERGY SECTORS

A. ARULRAJ
D. RAJA SEKARAN

SERIALS PUBLICATIONS
NEW DELHI (INDIA)

First Published – 2009

ISBN: 978-81-8387-219-5

Published by

SERIALS PUBLICATIONS
4830/24, Prahlad Street, Ansari Road
Darya Ganj, New Delhi-110002 (India)
Phone : 23245225 Fax : 91-11-23272135
E-mail: serials@satyam.net.in

Dedicated to

Our Parents

and

Management Gurus

CONTENTS

PREFACE

Power sector reforms as a part of economic reforms programme of government of India started in 1991. Crippling power scenario was a worry some national priority. The power sector reform became a national guide. The main stay thermal utilities are the fore runner for effecting major performances in this sector to save the sector. This book brings out early stages of power ventures, mid surge and slide, reform's sweep. As the demand and supply mismatch increased, technological change, reformative changes has come in handy to help the sector. The sector primarily depend on government. Various inherent and known problems were not addressed properly and not treated seriously. This situation had led to a huge loss accumulation that start paralyzed the sector.

Per-capita power consumption had increased two folds from 2001 to now. These fast changes have to cope up with power demand and supply positions. Governments were not able to succeed in correct direction. But the huge demand and supply gap start increased. A grim situation started crippling the sector and a unhealthiness that start ruining the health of plants. This book explains the treatment of technical, commercial, reformative combinations that start surfaced at least to address the pitfalls and later to serve as a tonic to a dying patient. Miracles started working through reforms. State Electricity Boards started slashing huge losses. Power theft that swallowed the sector are identified and start prevented at least in paper. It explains how the thermal utilities spring back to timely need in reforms process and nursed the sector.

The book elaborate past scenario, world scenario, post reforms developments in power sector, need for fast changes in all areas, competition need, customer satisfaction. This book exhibit the fire for power sector development and its strategy. It brings out analytical and statistical tools used to ascertain the role and success of thermal utilities performance in the development of a weak economical work and reformation process. The utilities helped privatisation process. The author's suggestions may help in the national priority sector's progress. This may open eyes on the pitfalls for its redressal. Finally the authors vision on national development through vital contribution of thermal utilities will pave ways for the sick sector.

We would like to express our sincere thanks to all faculties of Adikalamatha Institute of Management. We express my heartfelt thanks to Shri R. V. Shahi, Former Secretary of Power (GOI), Mr. S. C. Manocha-former Director Utility Powertech. Limited, Noida, and Ex. E. D. of Reliance Energy Limited, former Chief Executive Mr. N. P. Singh, Utility Powertech Limited, Noida and former Executive Director–NTPC, Mr. Mathew. P. M.–Deputy General Manager–HR, Utility Powertech Limited, Noida, and NTPC Limited, Mr. B. K. Bhattachrya, formerly Chief of Projects, TATA Projects, Secunderabad, for their management visions and support and encouragement for this work.

We sincerely express out gratitude for the support extended by the library officials of PMI–NTPC, Noida, IIM Ahemedabad, NPTC–Shaktinagar, Kawas, Jhanor, Vindhyachal, Dabhol, Reliance Power's Dhanu Electricity Board officials, Ministry of Power Officials, Reliance Energy Library Support, NTPC Noida Library Support, BHEL, Trichy Library, Er. Daniel Sagayaraj, Inspector Boilers (BHEL) for this work and IIM, Ahmedabad library and Er. E. Papan, Ex. GM BHEL, Tiruchirappalli.

Our sincere thanks to our office colleagues at Utility Powertech Limited, Noida, (a joint venture of Reliance Energy and NTPC) who have contributed their part by means of assistance and encouraging us to bring this work in a successful manner. Our heartfelt thanks to the TSSL, Trichy team for accomplishing this book work in a complete form.

Our heartfelt thanks to M/s Serials Publications, New Delhi who had rendered a major and magnificent job to bring out this book in a shortest time.

On home front, we must express my thanks to our beloved parents, Mrs. Reetha, Mrs. R. Shantha Rani, Mr. R. Dhayalan, Mr. R. Dhayalan Miss. R. Dhana Shri, Master A. Prabhkaran and Miss. A. Nikkitha who had helped at each and every stage of this work.

A. Arulraj & D. Raja Sekaran

1

NEED FOR REFORMS AND ROLE OF THERMAL POWER IN INDIA

INTRODUCTION

Power is one of the most important forms of Energy necessary for the economic development of nation. Power is a key input factor for ensuring and accentuating orderly socio-economic life to maximize welfare. Full utilization of other inputs and resources like man power, land (irrigation) and capital related resources rely on the easy and adequate available of Electricity. The draft 9th plan[1] rightly stated thus: "Electricity is the most versatile form of energy and provide infrastructure for economic development .It is a vital input for industry and agriculture: and is also of particular importance to developing rural sector. The future development of the country therefore will depend upon the rate of growth of Power generation capacity." The role of power sector is so influential that economists often establish a one to one correspondence between energy and economic development.

Power is the back bone of our economy and for any nation in the world for improving quality of life. Power sector accelerate growth of a nation's GDP. India's power sector has grown tremendously since Independence in 1947, with installed capacity rising 8% annually to more than 1, 24,287 MW by 2007. Despite this eighty-fold increase, per capita consumption of electricity is only 616 kWh per annum in 2006, much lower than world average of over 2,000 kWh. Indian's often compare their performance with China, which had lower level of development two decades ago, but today has more than double the per capita electricity consumption (and GDP) than India. Indian power sector had grown since 1897 to the present installed capacity of 1,32,329 MW as on 31.08.2007 (Ministry of Power). India maintains a low per-capita power consumption of 366 Kwh. in 2001 As compared to USA's 12441 kWh, Canada 15758 kWh, Japan 7405 kWh, China's 872 kWh and S. Korea's 4497 kWh 1998, as per CEA data (Central Electrical Agency, 2000). End users of power in the country are faced with shortage of power, frequent power cuts, voltage drops, frequency variations leading to severe daily life difficulties and economical loss to nation.

According to 16th Electric power survey[2] conducted by CEA which reveals recommendation for improving power generation in India are below: (i) Based on the projections made in the 16th Electric power survey conducted by CEA, over 1,00,000 MW additional generation capacity needs to be added by 2012 to bridge the gap of demand and supply, accordingly demand side management and additional capacity plan formed part. (ii) In view of high investment involved in the installation of new plant, urgent need for optimal utilization of existing generation capacity and to renovate and modernize (R&M) and life extension (LE) of existing old plants were considered to be most economical option but to obtain additional generation. (iii) Electrification of villages had seen as 77,897 villages still do not have access to electricity. (iv) In adequate power supply/Generation capacity leading to capacity expansions. (v) Optimum utilization of existing generation capacity, energy conservation. (vi) Increasing the efficiency of Power generations, (vii) Inadequate inter regional transmission links and development, and (viii) Electrification of Villages and households. On the other hand, Agricultural consumes almost one third of power in India, yet provides less than 5% of the revenues. Official estimates of the subsidy to agriculture are based on the difference between the tariff charged and the average cost of supply for all users. The book address a number of issues related to changing trends in electricity generation and its impact on power sector reforms in thermal utilities.

India's power sector is undergoing significant reforms, beginning in 1991, which are changing and diminishing the role of government, which functioned earlier as the near monopoly integrated utility. From being government departments, the State Electricity Boards (SEBs) are being unbundled, and the role of the government might be recast as the regulator and (last resort) financier, with operations in the hands of companies, especially private companies or privately controlled companies.

Indian power sector have four major sources, namely thermal, hydro, nuclear, non conventional. Of these sources Thermal accounts for 65%, Hydro power 26%, Nuclear 3%, and non conventional 6%. Thus forming a major source of power as Thermal source.

In thermal power there are three sub sources such as Coal/ lignite, Gas, diesel form resources of power. Of the total installed capacity of 1,32,329 MW as on 31.08.2007, coal based power accounts for 55% which is 72, 781 MW. 1, 00,000 MW additional generations capacity needs to be added by 2012 to bridge the gap of demand and supply. Country's total generation capacity as of November 2007 end stood at 1,38,251 MW. Two third of eleventh plan period capacity addition are based on coal plants and the 10 ultra mega plants are all coal based.

Other wise huge power shortage will cripple the country. India is abundant with coal resources. Import of quality coal for mixing for fuel efficiencyare gaining grounds. More than half of national power is generated from this sources and will start import of coal from coal rich countries. There is a growth rate of 6% in 2006-07 in domestic production of coal. Captive mining start supplying larger interest. Till September 2007 coal ministry had allotted 165 coal blocks and additional 15 blocks are being allotted by them.

Thus, the book addresses on the power sector reforms after 1991 in the context of thermal power. Economic reforms of the changing scenario necessitated power sector also to undergo reforms. The reforms in the Indian power sector, that were initiated in 1991 for mobilizing private sector for addition of power generation in installed capacity, renovation and modernization of plants, healthy performance orientation. Performances of these power utilities and their healthiness are important aspects of power sector, which is fast decoracting due to subsidies of state governments and poor status of State Electricity Boards (SEBs). The current thrust of reforms are on the generations, transmission and in distribution sector, reducing losses, privatization and increasing efficiency, computarisation.

Given India as a large nation, with 29 states, each states will and do behave differently. This dynamics, with increased competition between states is a positive sign in the reforms and development process. At the end of the day, India's reforms have gone a fair way towards the ingredients necessary to reaching the goals of increased access, efficiency, and viability, but they have not yet seriously and directly done so far.

Failure: These reforms are necessary but perhaps not sufficient; it will be the focus of enormous effort and expenditure by the government, funding agencies and companies in the thermal utilities. Country had failed in its capacity addition programme for three consecutive plan periods since VIII plan. This was mainly due to lack of monitoring and equipment supply problems as per MOP. Slow growth in all three (Hydro, Thermal, Nuclear) segments, resulted in power deficit increased in terms of peak availability 14.8% & total availability 8.4% in Jan. 2008 as per MOP.

Kandula Subramaniam (2004) explains that recognizing that electricity is the most convenient form of energy and an essential prerequisite for industrial development, the Indian Ii dustrial Commission, which reviewed the industrial development of the country from 1916-1918, stressed the importance of power development in the country and emphasized the need for a detailed hydroelectric survey to enable systematic development of water power resources. Responding to these recommendations, the central government instituted a survey of hydro-development potential. The period between the two World Wars witnessed the development of the Pykara, Mettur and Papanasanam hydroelectric projects in Madras (now Tamil Nadu); the Uhl river project in Punjab; the chain of power stations along the Ganga Canal in the United Provinces (now Uttar Pradesh); the Pallivasal Project in Travancore (now Kerala); and the expansion of the Sivasamudram Project in Mysore (now in Karnataka). In these states, grid systems began to emerge, as electricity from the hydroelectric projects was carried to distant load centers.

In the year 1989, the World Bank estimated that the requirements of the electricity sector in developing countries added up to US $100 billion per year. By contrast, only about US $20 billion was available from multilateral lending institutions, leaving a gap of about US $80 billion (Churchill and Saunders, 1989).

Requirement for additional power capacity to meet the projected demand for electricity formed the underlying basis of the power sector liberalization process, when the policy was announced in 1991. The growth of the Indian economy could not be held ransom for lack of electricity. Thus, in order to support an annual GDP growth rate of approximately 7 per cent, the power supply needed to increase by over 10 per cent annually (Ministry of power website, 2005).

Private investors entered the power sector as Independent Power Producers (IPPs). IPPs faced immediate worries about receiving payments in full and on time for the sale of power they made to the SEBs. Put simply, the private sector "will not take the risk of not being paid (Ahluwalia and Berry, 1996).

Any such risk factor was unacceptable because infrastructure projects yield relatively low returns and have long payback periods. Power plants are perceived to be commercially less profitable. Such projects had therefore been undertaken by the public sector. Private promoters faced difficulties when trying to obtain funding, as bankers were unlikely to agree to loans with maturity longer than three years, to match the tenure of their deposit liabilities. Even financial institutions (FIs) find it difficult to extend loans commensurate with the long payback periods of power projects (Antonette D' Sa, K. V. N. Murthy and A. K N. Reddy, 1999).

Heat Rate: A number of factors influence the heat of a generating plant, the choice of technology, size and fuel puts a limit on the thermal efficiency achievable. Other factors like vintage of the plant also play a role in determining the upper limit of efficiency. This heat rate is specific to each of the generating units and is known as the design heat rate. The design heat rate is supplied by the equipment manufacture. In practice, it has been found that the actual heat rate is higher that design heat rate. Studies carried out on selected plants at the state level have reported significant deviations of the actual heat rate from the designed one (Alagh, Shah and Shah, 1998). A number of factors like ownership of the plant, management practices, maintenance schedule, quality of fuel etc. lead to this difference between the actual and design heat rates.

Study that examine the policy incentive for adoption of energy efficient technologies, considering that these efficient technologies help in reducing emissions, are relatively fewer in number (Verhoef and Nijkamp, 1999).

The reforms have not only led to the entry of private players in generation, but have led to the mergence of various ownership structures, which were not present earlier. In addition to federal plants, state plants and plants owned by licenses, new structures like purely private initiatives and cooperative ventures between the government and private players have come. In addition, private organizations pooling resources for setting up generating plants and sharing the electricity over the state grid has become a unique feature of the state (P. R. Shukla, Debashish Biswas, Tirthankar Nag, Amee Yajnik, Thomas Heller and David G. Victor, 2004).

The driving forces behind the electricity sector reforms in developed and developing countries have been different. In developed countries, the main aim of the reforms has been to improve the performance of relatively efficient systems. In developing and transition countries, the burden of price subsidies, low service quality, low collection rates, high network losses, and poor service coverage have meant that many governments are no longer willing or able to support the existing arrangements (Newbery, 2002; Joskow, 1998). The reforms have sought to transform the state-owned and centralized electricity sectors into decentralized, market-oriented industries with private sector participation, competition in generation and supply businesses, and regulation of natural monopoly activities. In order for the decentralized industry to function, the reform and regulatory design must establish appropriate structural, institutional, and operational framework. The main steps of a systematic reform are to (i) restructure the sector, (ii) establish regulatory authorities, (iii) organize markets for generation, (iv) regulate transmission and distribution networks, (v) privatize existing assets and promote new investments, and (vi) allow for cost-reflective electricity tariffs (Newbery, 2002; Joskow, 1998).

Electricity sector reforms in developing countries have taken place within diverse political, economic, and structural contexts. In addition, many reforms were initiated at a time when the international experience with such initiatives was limited. Consequently, the reforms have taken a variety of forms and followed different paths (Bacon and Besant-Jones, 2001; Millan *et al.*, 2001). Within this background, it is perhaps not surprising that many reforms have encountered unexpected problems and the degree to which they have achieved their goals varies across the countries (Fischer and Serra, 2000).

The international experience with electricity sector reform in developing countries has shown that achieving workable reforms is considerably more complicated than anticipated. Electricity systems in developing countries vary considerably with regard to size, structure, and resource mix. In addition, many of these countries are constrained by institutional endowment of their political and economic systems and lack of human resources with regulatory skills and experience (Stern, 2000).

It is generally recognized that regulatory design and implementation strategy should take the specific characteristics of the sector in question into consideration. However, the reform models adopted have not always fitted the sectors of these countries and many reforms have encountered unexpected problems and unintended outcomes. The experience has shown that regulatory design is crucial to success and failure of reforms (IADB, 2001).

PRIVATE SECTOR PARTICIPATION IN ELECTRICITY SECTORS OF DEVELOPING COUNTRIES

Private sector participation is arguably the most important element of electricity sector reforms. For many reforming countries faced with increasing burden of capital requirements for expansion of publicly owned electricity systems, private participation is an alternative source for securing the much-needed investments in the sector. In addition, to the extent that the required public expenditures are financed through taxation, the marginal cost of public funding in terms of the associated dead-weight loss constitutes an added social welfare loss. In developing countries, the magnitude of such losses can be significantly higher than 1. For example, the dead-weight loss in Malaysia, Philippines, and Thailand have been estimated at 1.2, 2.5, and 1.2-1.5 respectively (Beato and Laffont, 2002; World Bank, 1997).

Private ownership together with competition (and incentive-regulated networks) is expected to result in cost efficiency, lower prices, reduced system losses, and improved revenue collection (Newbery, 2002).

Electricity supply industry is no longer viewed as a vertically integrated natural monopoly activity. Rather, the industry is regarded as separate but inter-related activities with distinctive economic characteristics. The new view of the electricity sector is that the generation and supply activities are potentially competitive while the transmission and distribution activities can, exhibiting natural monopoly characteristics, be subject to incentive-based regulation. The main focus of electricity sector reforms has been on liberalization of electricity generation. However, many reforming countries have experienced difficulties in enforcing effective competition in this market. It has now become evident that regulatory design is crucial for achieving effective competition in the sector. Lack of competition results in market power to existing actors that (i) reduces pressure on cost-saving efforts, (ii) limits consumer choice, (iii) distorts investments in new generation capacity, and (iv) prevents new entries. Several conditions can lead to lack of real competition and market power. First, in the restructuring phase, the reform must ensure that existing generation resources are split into a sufficient number of potentially competitive units. The main issues here are to avoid establishment of dominant firms and ensure a balanced resource mix among the competing firms while taking the size of the sector into account. In the UK, the problem of market power became apparent shortly after the reform. It was only after a lengthy process of new entries by IPPs and forced divestiture of generation capacity by the incumbents that a more competitive market was achieved (Newbery, 1999).

Economic theory suggests that cost-reflective prices result in net social welfare gain. This implies that the welfare economic gains by those who benefit from lower prices exceed the welfare losses incurred by those who stand to lose from price increases. However, without public interference no automatic transfer from gainers to (targeted) losers will take place to compensate the latter. Mapping and measurement of distributional aspects of tariff adjustments is an inherently complex task (Chang, 1997).

In some circumstances a price increase to efficient levels may also be socially defendable. For example, in countries with very low rates of access the service is often only available to richer consumers. Therefore, a rate increase that eliminates the system's deficit financed by the whole population and frees resources for improving access to others can be justifiable. For example, two important questions are the intensity and distribution of gains and losses across different groups of consumers. Welfare losses to some disadvantaged consumer groups can be much larger than the benefits accrued to many gainers. Also, in poorer countries losers generally constitute a very large portion of the population while the number of gainers can be far fewer. In addition, the higher the level of existing subsidies is the more noticeable are the distributional impacts of tariff re-balancing. There is considerable scope for efficiency improvement in distribution utilities. At the same time, tariff adjustments can play an important role in financial health of electricity distribution utilities and their ability in achieving these efficiencies. Price adjustments are therefore closely linked to the issue of privatization of distribution utilities. However, in many developing countries, the need for subsidies will be present for the foreseeable future. The important issue is to design subsidy schemes that address undesirable social impacts while limiting price distortions and adverse impacts on the economic efficiency of reforms. Some subsidy schemes can be either very costly as they also tend to benefit ineligible consumer groups (inclusivity issue), or that they do not reach the targeted groups (exclusivity issue). Inadequate attention to the distributional implications of price increases can severely affect the progress of the reform process. In India, the state of Orissa restructured and privatized the distribution companies. Despite substantial and politically difficult tariff increases (11% in 1997, 9.3% in 1998, and 4.5% in 2000), the privatized utilities experience severe financial difficulties. The companies are unable and lack incentives to invest in efficiency improvement measures (TND, 2002; Business Line 2001).

The experience form the past two decades has shown that achieving sustainable private participation and market-oriented electricity sector reforms are more complex than initially anticipated. The cases of the UK and California have demonstrated that even in developed economies reforms encounter problems and exhibit intended consequences. Developing countries have had to reform technically and financially less efficient systems with less developed private sectors, weak economic and political institutions, and shortage of human resources and regulatory experience (Tooraj Jamasb, 2002).

In developing countries, the experience with electric power markets has been quite different. In the late 1980s and early 1990s, all of the largest developing countries, along with many of the smaller developing countries and nations in transition in the midst of rapid economic growth, announced plans to restructure their power systems. In the years since, most actually attempted some market-inspired restructuring. Yet the electric power systems that are in place today bear little resemblance to the theoretical market-oriented ideal (Thomas C. Heller and David G. Victor, 2003).

Thomas C. Heller and David G. Victor, (2003) focus on five countries—Brazil, China, India, Mexico and South Africa. Their study reveals that the difficulty of reform stems from the nature of state-centered systems—their financing, the many ancillaries social functions that are performed by SOEs, and their systems of governance and control. Those attributes, we suggest, lead to predictable patterns of behavior in the reform process. We suggest that they will explain why efficiency alone is unlikely to be an adequate motivator for reform. They also suggest that the process of reform is likely to become intertwined with reforms in other aspects of the economy, such as fuel markets. Financing, in particular, would appear to be the key variable.

During more than fifty years, the Brazilian power industry was developed by state owned companies, operating under a cost plus tariff regime. During this long period, the interconnection of the regional grids was explored to improve economic and technical performance of the system. The large availability of domestic fiscal funds and international soft loans has been a key element in the successful trajectory of the industry in this period. However, the interconnection had a corollary: centralization of generation and transmission in the federal government companies, a process that found opposition in the state (provincial) companies. The reaction to this process came in the 1980´s, when successive unsuccessful macroeconomic stabilization produced a disastrous economic effect in the electricity companies. Until 1998, it seemed that the reform was a success but looking from today the reform is a complete failure. This shift in perception is the result of the fact that at the beginning of the process, the fundamental regulatory flaws of the reform have had no time to show up their detrimental effect. These flaws became visible during the rationing period. Chiefly among these problems are the need of mitigation risk mechanisms for private investors and the regulatory interlink between the natural gas industry and the power industry. In other words, the reform was unable to generate the climate for private investment, particularly in thermal power plants. The new government is promising radical changes in the industry but statization is not in its agenda so far. However, the new government offer signals that it will not favor competition and intends to roll back to government a substantial role in the decision making process, particularly as far as investments is concerned (Adilson de Oliveira, 2003).

India's energy sector revolves around State Electricity Boards (SEBs), which own and operate vertically integrated monopolies—in generation, transmission and distribution. Although SEBs were guaranteed a rate of return on their investments, in practice all have sustained heavy losses that the state and central governments in India are increasingly incapable of bearing. Large losses in the system are due especially to theft as well as to a transmission and distribution system that is not engineered for optimal reduction in transmission losses. In addition, the tariff-setting mechanism has been politicized and structured in ways that, in effect, subsidize agricultural consumption; now that agricultural interests favoring cheap electricity (used mainly for pumping water) are entrenched, the challenges to reformers have mounted.

Reform started in India in 1991 with a series of "fast-track" projects aimed to attract new generation capacity. Conceived at a time of broader reforms in India that were designed to open capital and trade to foreign competition, these Independent Power Producers (IPPs) were expected to operate on the basis of long-term power purchase agreements (PPAs) with highly profitable tariff structures that were intended to attract outside investors. Among these were the notorious Dabhol project, plagued by negotiations that lacked transparency and guaranteed tariff that was extremely high—leading to allegations of corruption that are a shadow over the project even today. In the end, few "fast track" projects were actually built, and the focus on generation has proved to be misplaced. The root cause of India's power crisis is the lack of profitability in the final distribution and billing for power—a problem that requires a focus on the SEBs (Rahul Tongia, 2003).

India needs to focus on capacity expansion, policy-makers should look more carefully at all the options available at the local level. In particular, the possibility to franchise rural service areas to private companies deserved to be investigated more thoroughly. With regard to the environment as well, new options to modernize the sector and provide environmental protection could be explored, specifically by: (1) exploring gas as an end-use energy alternative (use for cooking, etc. rather than for thermal energy production); (2) exploring decentralized generation; (3) increased efforts to support demand-side management; and (4) investigate carbon subsidies. However, all of these options depend critically on new financing and significant regulatory capacity. The third speaker provided an investor perspective on IPPs. First off, he mentioned that there is an unprecedented melt-down currently going on in the share values of power developers. Rather than the poster child, energy developers and utility companies in general are now seen as risky investments by the financial capital markets. A previous speaker cited economic force majeure as the major cause for the investment slump, but investors usually regard these factors as part of the normal business cycle. And while it is true that power purchase agreements are changing, this speaker commented that investors saw PPAs only as stepping stones to a competitive market and not as solutions that were sustainable in the long run (The Political Economy of Power Market Reform, Conference Report, Stand Ford University, 2003). Arul Raj & Raja Sekaran (2006) explained that due to shortages of power generation and supply in India export of power from neighboring countries which has huge hydro potentials that can be developed along with neighboring countries in the BIMSTEC nations, (Bay of Bengal League of Nations for economic co-operation) energy co-operation policy. These countries include Bangladesh, India, Myanmar, Srilanka, Thailand, Nepal and Bhutan. A novel method of capacity addition, as to have the economies of large scale unit as a promotional measure envisaged in the form of ultra mega power projects (UMPP) by the Government.

CONCLUSION

Financial Performances and operational performance of the thermal utilities have improved way beyond during power sector reforms period and paved ways for loss reduction and sector development. Plant load factors have creditably improved to set standards and recover cost of installations. Capacity addition programmers are achieved but not very much. Nonetheless, they are a step in the right direction, ending years of government control, monopoly and mindset. The government corporations-state electricity boards are performing worse than private companies, but having all of them competing to perform better and it will only help the consumer and the main stay thermal utilities as a whole since coal fired stations are the highest share of 43% of all fresh additions in the years 1990-2007 period. Sizable capacity increase in future are fuelled by coal although raise in gas, hydro and renewable sources. Coal Division Document 2025 project an estimated demand in 2024-25 by 1267 Million tons from the present domestic production of 425 Million tons. Thermal power shares had increased to 73% from 59% in the last two years. There is a need for coordination between generation utilities, transmission utility, distribution utility, and consumers in India for further better performance for a bright India. The book elaborate these perspectives.

Notes

1. The Draft 9th Plan-Planning Commissions GOI Draft Proposal 1996.
2. 16th Electric power Survey Report by CEA-2001 for power projections.

2

POWER SECTOR IN INDIA: AN OVERVIEW

INTRODUCTION

PRAYAS Energy Group (PEG) is working on electricity sector policies with an aim to protect and promote public interest in general and the interests of disadvantaged sections of society in particular. PEG have reviewed electricity sector developments in India in an impartial manner.

The PEG report on India's historical review put it, that India has continued to achieve significant growth in the last 50 years. In the Constitution of India, electricity is a concurrent subject handled by both the central and state governments. Following this, the report say, a network of institutions are established at the state and central levels. The State Electricity Boards (SEBs) are created at the state level. These boards, owned by the state government, were semi-autonomous bodies according to the law. They were entrusted with the responsibilities of electricity generation and its supply in the state, while remaining within the broad legal and policy frameworks designed by the state and central governments. Few landmark developments like Creation of Central Electricity Authority (CEA), NTPC Limited (NTPC), National Hydro Power Corporation (NHPC), etc to add thrust to the development to the electricity sector under the jurisdiction of central government. The report say that Generation capacity increased to 94,000 MW in 1998-99, Number of consumers grown to 7.85 crores, Agricultural pumps connected upto 1998-99 are 1.20 crores, length of Network 0.5 crore kilometers, number of villages connected to power in 1998-99 are more than 5 lakhs and a per capita annual consumption rose to 340 kWh or units. This is a commendable growth of the Indian electricity sector as a result of four broad policies. The *first* policy was government's *ownership and supply* of capital from central and state budgets'. For many years, about one fourth to one fifth of the total plan allocation from the central and state governments' budgets was directed to the electricity sector. The *second* major policy was development of *'centralized electricity* supply system' and of 'regional and national electricity grids'. Under this policy , plants like Bhakra–Nangal, Singroli, and Koyana as well as the five regional grids were developed, *Third* major policy put *'self-reliance in technology and fuels'*, accordingly autonomous but government-owned companies like Bharath Heavy Electrical Limited (BHEL) were created to manufacture technological capabilities in the power sector. Importance laid on use of the available energy sources like coal and hydro source in the country. *Fourth* policy of *'cross subsidy'* or subsidy from within the sector was adopted with an objective to provide power at affordable rates to the deprived sections of society, especially farmers from backward, rural, and tribal regions. Higher rates of electricity were charged more than the average cost of supplying electricity and the surplus thus created was used to provide electricity at lower rates than the average cost of supply to the deprived sections of the society.

CRISIS IN THE SECTOR

PEG report states that above four policies have developed many distortions and perversions in the functions of the electricity boards and led to three crises in Performance, Finance and credibility. In Performance low efficiencies, financial disorder, and lethargic administration and that these distortions in the functioning of SEBs are felt beyond improvement.

In the financial crisis that electricity boards are facing a severe financial crunch due to stagnant revenues, increasing expenditure, and increasing arrears. The state and central governments that, until now, provided funds to SEBs are also equally cash-strapped. International financial institutions, fed up with the functional anarchy in the SEBs gradually stopped providing them funds. In such a situation, it has become impossible for the SEBs to raise capital to construct generation plants and transmission and distribution (T&D) systems in order to fulfill the increasing demand for electricity. On credibility, functional anarchy in the administration of the SEBs has reached such a level that the electricity boards have completely lost credibility in the eyes of consumers and common citizens. Mis-understanding on the part of the sector–leaders has been largely responsible for further aggravating the crisis.

ECONOMIC REFORM

PEG report explains that the electricity sector already in three-dimensional crisis, but all the actors were looking for a solution for the financial crisis alone. The economic reforms got underway gave an easy way out to resolve the financial crisis in the electricity sector, by inviting private capital and opening the closed gates of the sector to foreign investors. In the electricity sector, many fundamental and comprehensive changes were made in the laws governing the electricity sector, in the institutional structure, and in major policies and procedures. Three major changes are listed as illustrations. The first is the changes made in the Electricity Supply Act of 1948 to allow entry of private capital. Second, various policies and procedures were changed which resulted in severely curtailing the role and authority of the Central Electricity Authority (CEA) and were delegated to the electricity boards, state governments, and newly created independent regulatory commissions. Third, the fuel policy was changed in a major way to allow import of oil and gas. Beginning of the era of economic reforms-liberalization, globalization and privatization–gained ground the electricity sector.

First step in the 'reform' of the electricity sector was to allow the entry of electricity generating plants owned by private parties. To facilitate this, central and state governments made comprehensive and fundamental changes in various crucial aspects of governance such as laws, policies, institutions, and even in procedures. These changes had three important features. First, 'obstacles' such as permits and quotas were removed and SEBs and state governments were allowed to seek and sanction generation projects by dealing directly with private firms. These private firms owning power generation plants are known in the sector as Independent Power Produces (or IPPs). Second, state governments and SEBs signed agreements with the private parties that gave guarantees of high levels of fixed revenue to the private firms, while shouldering almost the entire burden of various risks. Third, the central government started providing different types of assistance for these efforts by SEBs. For example, the central government gave the special status of 'fast track projects' to the eight IPP projects in different states and also gave counter-guarantees to these projects in the event of non-payments by the state governments and SEBs. The report explained that for more than 90000 MW capacity MOU's were signed but only 8000-9000 MW would come on line in 2005. The IPP's were faced with high project cost, fuel supply, not getting timely payment from sick SEB's for the power purchased. The report said that undue secrecy on sanctioning projects and fuel, payment fronts created distrust and suspicion was created on the projects and led to political, legal controversies.

INFLUENCE OF THE WORLD BANK (WB)

As per PEG report that the World Bank as a lender and development institution had pressed for SEB's improved function and now have resorted to canceling six loans and resorted to rigid position that privatization be the precondition for any financial assistance to SEB's. WB Way to save the electricity sector is to replace the SEBs by privately owned electricity companies. Failure to reform SEBs and economic developments at the international level has played a key role in shaping this position of the World Bank. Fall of the Soviet Union, the USA and other nations established a clear domination in the 'unipolar' world and the manifestations of this dominance is the wide-scale acceptance of liberalization, privatization, and globalization. World Bank and the international Monetary Fund have been the chief proponents and leading campaigners for LPG.

The PEG explains that the pressure due to the financial crisis faced by SEBs–and, on the other side, the World Bank, despite its previous bad experience of state governments, willing to enter into (possibly a temporary) compromise with the state governments for promoting its 'LPG' agenda. The first manifestation of this emerging coalition was in the form of the World Bank promoted comprehensive electricity sector reforms in Orissa. Therefore, the broad model adopted by the World Bank for electricity sector reforms in India is called the 'Orissa Model'. Due to the immense political and economic power of this new coalition, the Orissa Model spread very fast in many parts of the country. Following this model, the Asian Development Bank (ADB) also started working with state governments in Gujarat and Madhya Pradesh for structural reforms in the electricity sectors of these states. By now, more than half the states in the country have adopted some variant of the Orissa Model for electricity sector reforms.

The report further explains that under the leadership of the late Mr. Kumaramangalam, who advocated reforms in the electricity sector, the Union Ministry of Power pushed for the spread of the World Bank's Orissa Model in other states. In 1997, the first draft of the central bill for electricity sector reform was introduced. In 1998, after deleting these unacceptable provisions, this bill was passed and enacted as the "Electricity Regulatory

Commissions' (ERC) Act, 1998" The Act leaves two important issues to the discretion of the state governments-the timing of formation of the state regulatory commissions and the powers to be delegated to these commissions. However, the Act had provisions to give the boards substantial powers fro determining electricity tariff to these commissions.

The report put it that the states of Orissa and Andhra Pradesh were in the forefront in implementing the World Bank's mode. Different issues and problems have emerged in implementation. Other states can and need to learn many lessons from their experience. One of the most important lessons that emerged from Orissa's experience is that the process of privatization of SEBs is a time-consuming process and the projected gains of privatization do not accrue fast. In Orissa, four years after the privatization process has been initiated, the transmission and distribution (T&D) losses have remained at the same level (more than 40%). As a result, the independent regulators did not find it appropriate to allow the tariff hike to be as high as demanded by the utility. As a cumulative effect of the high T&D losses and low level of tariff hike, there was a serious financial crisis in Orissa's electricity sector. To save their beloved progeny from this crisis, the World Bank and the Government of India had to chip in a few hundred crores of rupees to tide over unexpected cash flow problems.

In the process of privatization the study explain that, various structural flaws have also crept into Orissa's electricity sector. For instance, while privatizing the distribution system, the state was divided into four regions. Distribution companies in three regions were sold to a single company, viz., Bombay Suburban Electricity Supply (BSES). Neither BSES nor any other company was ready to buy the government–owned distribution company in the fourth or central region. This was because that state government had already signed as agreement (called an escrow agreement) with an American electricity generating company called AES Inc. giving the company the first right to access revenue collected from the central region. As a result, nobody was ready to purchase the distributing company in the central region. In the end, the state government and the World Bank were compelled to sell the distributing company in the central region to AES.

Orissa and Andhra reforms are explained in the study that, to improve Orissa's electricity sector, the World Bank wanted to break the monopoly of the state and to bring in competition. In the process, it ended up thrusting private monopolies on the people of Orissa. In electricity distribution, BSES has established a horizontal monopoly over three-fourths of the state, while in the central region, AES has established a vertical monopoly in electricity generation and distribution in the region. The people of Orissa will have to suffer due to these structural flaws at least for the next few decades. This clearly demonstrates that the World Bank and state governments–who were initially promising that they would take utmost care to avoid unintended bad effects of privatization–in reality, are ready to compromise on their own principles for implementation of their agenda of privatization. This also indicates that, in reality, all the promises–made by supporters of privatization-of protecting public interest (through various safeguards), would be blatantly ignored to fulfill the objective of privatization of the sector, if necessary. In addition to such structural flaws, another crushing burden was placed on the common people of Orissa during the process of privatization. While selling the state–owned utilities to private firms, with the intention of earning maximum possible profiles, the state government imposed an additional burden of Rs. 2,000 crore on electricity consumers in Orissa by increasing the sale value of the SEB through the mechanism of 'asset re-evaluation'. At lest for the next two decades, the private companies will charge heavy tariff for recovering this additional price they paid to the government. Unfortunately, very little of this additional revenue was used to improve the sector.

The review explains that in Andhra Pradesh, the then Chief minister and the World Bank started implementing reforms in a high-handed manner. Neither the Chief Minister nor the World

Bank experienced any need to take the other 'stakeholders' into confidence. The opposition parties, people's organizations, consumer organizations, employees' unions, and farmers' organization were systematically and purposely kept at bay. This coalition started using various pressure tactics for implementing electricity reforms. Members of the state legislative assembly (known as MLAs) belonging to opposition parties were summarily suspended when the electricity reform bill was passes in the state legislative assembly.

Unfortunately, the independent state regulatory commission created by the World Bank, which is being seen by some as a pawn in the hands of this coalition, seems to have lost all its credibility with the people and civil society organizations in the state. The commission is accused of neglecting even the mandatory provisions related to transparency, accountability, and participation in its proceedings. It is alleged that the commission allowed to keep secret the documents highlighting irregularities on the part of the state government and thus helped the politicians who were benefited by these deals. As a result, when the regulatory commission-that apparently has lost its credibility in the eyes of people–declared a tariff hike following the World Bank's dictates, the entire state erupted in political turmoil. Despite all these developments, the World Bank, the American and other foreign governments, and some mainstream economists in India have been complaining that the state governments in India are not carrying out electricity sector reforms (which, means privatization) with the necessary vigor and speed.

GROWTH AND UTILIZATION OF POWER SECTOR

V. S. Kulkarni (2004) empirically explained that our progress in the power sector has been remarkable in generation on one hand but at the same time the sector is engrossed with a number of problems. The growth profile shows that the installed generation capacity of power increased from the mere 1362 MW in December 1947 to 106800 MW in December 2002 of which 72% are thermal

25% hydro and 3% nuclear power. Between 1970-71 and 1999-2000 installed capacity has grown at an annual rate of 6.9% and generated power rose from 55.8 billion KWh to 515.3 billion KWh.

In first decade of post reform period the installed capacity has gone up from 4973MW to 9857MW between 1991-2000. There has been marked increase in domestic sector from 9% to 22% due consumer demand. The shift in consumer pattern is well explained by the study.

Pattern of utilization of power are as below (in percentage) as per GOI Economic Survey 2001-02 TATA Services Ltd., Statistical outline of India 2000-01.

Table 1
Shift in Consumer Pattern

End-User	*1970-71*	*2001-2002*
Industry	68	35
Agriculture	10	29
Railway Traction	3	3
Public Lighting	10	11
Domestic Use	9	22
Total	100	100

He explain that power utilization of Industry has come down from 68% to 35% between 1970-71 and 2001-02 attributed to a number of industrial units have setup their own power plants instead of depending on public utilities. On the other hand there has been an increase in agriculture sector from 10% to 29% largely on account of irrigation and government subsidies.

GROWTH OF POWER UTILITIES AN OVERVIEW

Power Sector's Preliminary Stages: Electricity supply begin in India by 1897 by constructing first Indian hydro power station in West Bengal at Darjeeling with an installed capacity of 130 KW, Calcutta Electricity Supply Corporation (CESC) Limited, Which was registered in London are granted License for 21 years to illuminate and power the area of present Kolkotta and supplied power at one rupee per KWh on par with London's tariff In 1899 1 MW then Calcutta installed thermal power plant. Mumbai see light with the private company built urban power supply. Followed by this city of Ahmedabad, Surat is illuminated with power.

In 1902, 6x720 KW Shivasamudram hydro power station in Mysore state established along with worlds longest transmission line upto Kolar Gold Fields in Karnataka (Sankar and Ramachandra 2000)

From 1914 to 1926 Khapoli hydro power station are established at 48 MW and later increased to 72 MW. In 1924 Bhivpuri 72 MW hydro station and in 1927 Bhira hydro station of 110 MW established by Tata Electric company and laid power foundations in India. The installations upto 1947 are at 1362 MW.

State Electricity Boards/Utlities: Majority of private companies are amalgamated to after independence into state owned companies. In 1948 Electricity Supply Act, formed on the lines of 1926 British Law. This has created state owned State electricity boards (SEB's) with a responsibility to generate, transmit and distribute power (Chokroun 2001). The SEB's also acquired private owned services as their license expired gradually. They start invest on generations. In 1950 as the Indian constitution gave power in the concurrent list of the federal Indian set up. In this set up central authority override the states. The social elements of employment profits ploughed back as investment for growth of the field. SEB's are interconnected by grids to enhance system reliability instead of earlier isolated services. These SEB's are legally autonomous bodies of a state for the power management and are required to generate a profit along with 3% rate of return on the net value of its assets. Inreality the SEB's internal accruals are insufficient for growth and depend on state for grants subsidies, soft loans.

Public Companies of Central/Central Utilities

The central government to assist states and to augment support establish a number of publicly owned companies for generation of power and transmission in many states during nationalization policy of the then prime minister Mrs. Indra Gandhi in 1975 two companies namely, National Hydro Power Corporation (NHPC) and National Thermal Power Corporation (NTPC). The NHPC formed Hydro projects had the difficulty in project delays and to be consulted with internal water resources, environmental, societal difficulties. NHPC had grown today with 32000 MW as on 2006 (CEA Data). Many monsoon seasons' failure give a thought to develop to exploit Indian coal reserve in large power plants in many states, thus the formation of NTPC incorporated in 1975 and had grown as the world's 6th largest power capacity with 24000 MW above. Both NTPC and NHPC are maintaining world standards and earn profit more than 12% returns on capital. The thermal major has aggressive expansion in future with power trading, consultancy, distribution, etc.. To cover loss making SEP's.

Transmission of Power/Transmission Utilities

Transmissions of Power to feed SEB systems are facilitated with five regional synchronous grids with own regional electricity board and dispatch centers are established by NTPC to create an integrated national grid. In 1980 the government separated transmission assets from NTPC and formed Power Grid Corporation of India (PGCIL) with responsibilities for all interstate power transfers. Today PGCIL grown as largest transmission company in the world with more than 40,000 KM of transmission lines linking one third of India's generating capacity to load. Presently the company profits more than 28% (Net Profit) which is very high for a transmission company.

The combined installation of state and central power utilities along with private utilities had crossed 1 Lakh MW in 1998. The supply of growth rate has not kept pace with the demand growth. India's lowest in the world per capita consumption is only 3.4% of generation but having 17% of world population in 1998.

Distrbution of Power/Distribution Utilities

Causes of Financial Sickness: Distribution system in various states have grown in an unplanned and haphazard manner without optimal network size and location of substations and adequacy of back up transmission. This has resulted in very high power losses, poor quality and reliability of power supply and in bad financial condition of SEB's. The status of SEB's are reasonably good upto 1975 and start detoriated due to poor metering, heavy subsidies and free power schemes to consumers, highly subsidized flat rate tariff for agriculture, single point unmetered domestic customerism, persistence theft of power leading to more than 40% of power generated leading to heavy transmission and distribution losses, high level of cross subsidy forced industry to switch over to captive power generation, lager manpower usage in the distribution leading to higher cost, Lesser billing collection leading to revenue losses.

The energy lost during voltage transmission as well as transportation through transmission and distributions are technical losses. Power lost by theft and pilferage are non technical or commercial losses. In the Indian power system the billed portion of power over generated power is 55% only.

Energy Loss Reduction Strategy

Improvement of system and reduction of losses could be effectively achieved by conducting energy audit. The energy audit is attempted starting with areas known for excessive losses.

- Access overall energy loss in a specified area (sub-station or distribution center).
- Identify system elements passing excessive loss.
- Identify and establish causes of excessive loss (Technical or non-Technical factor).

Use of user friendly software packages to create a comprehensive system database and study a number of development options for evolving least cost plants that can ensure economy, efficiency and security of supply.

Privatization of Distribution System

About 75 to 80% of total work forces are involved in this area larger low voltage lines and cables are involved, larger energy tapping points are involved for a better performance.

- Load balancing and line loss management within reasonable limits to be managed.
- Checking unauthorized load and penalty for tamper of energy.
- Tackling pilferage, breakdowns, and voltage fluctuations.
- Ensuring safety and accuracy of millions of Energymeter.
- Prompt and timely attention on consumer complaints in breakdown, over billing, wrong meter reading, making alternative arrangement for uninterrupted power supply.
- Improved operating, capital investment deficiency to make technical financial and managerial competency.

Power Sector Reforms since 1991–A Chronology

Dr. Kulkarni in his Policies and Perspectives in Post-Reforms Period has explained that with the introduction of new economic reforms through New Economic Policy (July 1991) the government is looking for growth strategy through economic liberalization and privatization to achieve globalization. As part of this reform process the government has introduced sweeping reforms in industry, trade, and banking and in telecommunication sector. The reforms in power sector are part and parcel of the governments' privatization process for a shift in pattern from state controlled to private, competitive, and independent regulations. Major reforms pursued by successive governments since 1991 are aimed at decontrol and de-bureaucratizing the power sector.

The Electricity Laws (Amendment) Act, 1991: Which replaces the earlier Indian Electricity Act, 1910 and Electricity Act, 1948, and aims at, (a) allowing private sector to set up coal, gas or liquid fuel-based thermal, hydel and solar/wind projects of any size (b) allowing foreign investors upto 100% ownership of power projects subject to government approval. (c) setting new price structure; (d) new power projects are eligible for a 5 year tax holiday and (e) Duties on import of equipments for projects have been considerably slashed.

1992-97: A high level team visits U.S, Europe and Japan to invite foreign investment in power sector. *1992-97 eight* projects were given "Fast Track" status in the form of Sovereign guarantees from central government. Seven projects reached financial closure. They are-Dabhol/Enron, Bhadravati/IS Pact, Vishakapatnam/ Hinduja, Jegurupadu/GVK, IB Valley/AEs, Neyvel/CMS and Mangalore/Cogentrix.

1995-96: Orissa Electricity Reform Act and establishment of OERC (Orissa Electricity Regulatory Commission) to monitor and control generation and distribution are introduced.

1996: At the Chief Ministers Conference and its Common Minimum Action Plan for power recommendation to create state and central regulatory commissions–(SERC, CERC) with the CERC having responsibilities to promote competition, efficient and economy to improve quality of power supply, to promote investment, to advice government on removal of institutional barriers, to safe guard consumer interest etc. The SERC are given autonomy in their respective states in the matters like Licensing, planning determination of retail tariff, cross-subsidization, fuel adjustment charges, package of incentives and dis-incentives, inventory control, R&D, Project Management etc.

1997: Liquid policy introduced for easy naptha allocations to private sector investors.

1998: Mega power policy for the construction of and operation of 500 MW and above hydro power projects and at least 1000 MW of thermal power projects.

The Electricity Laws (Amendment) Act, 1998 and Electricity Regulatory Commission ordinance Notification–provides for (a) creation of central transmission utility, (b) Establishment of CERC and SERc (c) Rationalization of electricity tariff. (d) Policies on subsidies. (e) Promotion of efficient and environmentally suitable policies. Accordingly Power Grid notified as Central Transmission utility, Haryana Electricity Reforms Act, World Bank loan of US $ 600 million Adaptable Programmed Loans (APL), U.K's (DFD) technical cooperation grant of 15 million pound for reforms are effected.

1999: Andhra Pradesh and Karnataka Electricity Reforms Act, World Bank loan of US $ 210 millions under APL, DFID'S UK grant of 28 millions as technical cooperation, and CIDA (Canada) technical assistance of Canadian $ 4 millions are effected.

2000: Power Ministers Conference and Electricity Bill 2000 (Draft) that focused on (a) Functional desegregation of generation, transmission, distribution to create independent profit centers and accountability, (b) Reorganization and restructuring of SEB's with a freedom to state government to prepare model/plan, phasing and sequencing of restructuring, (c) States to decide extent and nature of privatization, (d) Central and State Regulatory Commissions on the lines of Electricity Commission Act 1998, (e) SERC need to function for equitable supply of electricity to rural areas and to weaker sections, (f) Stringent provisions to minimize theft and misuse.

2000-01: Government initiation of Accelerated Power Development Programme (APDP) to provide financial assistance to the states for renovation and modernization of Hydro and thermal power plants and to strengthen transmission and distribution network.

2001: (a) Under the chairmanship of Montek Singh Ahluwalia an expert group formed to recommend one time settlement of past dues by SEB's to Central power sector utilities (CPSU's) and State Power Utilities (SPU's) and the main recommendations include tapping of market borrowing and private investments to finance expansion and modernization plans in the sector, waiver of loans to SEB's and central assistance for bridging revenue gaps, meeting the costs of the workforce rationalization etc. (b) Energy Conservation Act 2001, to promote conservation of energy and to balance demand and supply to avoid power shortages, (c) The Electricity Bill 2001, introduced in in parliament in August 2001 that replaces earlier three Acts and focuses on rural electrification, management of rural power by local authorities etc.

2002: The government initiate and focuses on an objective of providing "Power for All by 2012" with a "Mission 2012" in this direction.

2003: The Electricity Act, 2003 has come into force from June 2003 with a roadmap for the electrical industry, and hopes to speed up economic reforms. It is an outcome of a series of commissions, national and international consultancy studies, seminars and conferences in the last three years, since the Draft bill 2000 was outlined.

2004: Accelerated Electrification of One Lakhs Villages and One Crore house holds by merging rural electrification and Kutir Jyothi programme for providing 40% capital subsidy for rural electrification projects and the balance as loan assistance on soft terms from Rural Electrification Corporation.

2005 : (a) National Electricity Policy are notified in Feb. 2005 aiming at accelerated development of the sector, providing power supply to all areas with reliability and quality, protecting interest of the customers and other stack holders and at financial turn around and commercial viability of the sector. (b) The Electricity (Ammendment) Bill 2005 has been introduced in Loksabha on 23.12.2005. (c) Notification of Electricity Rules, 2005 that include consumer redressal forum, tariff of generating companies.

2006: (a) National Tariff policy under Electricity Act, 2003 Notified in Jan 2006. (b) Central government made the Appellate Tribunal for Electricity for hearing appeals against order of Regulatory Commissions/Adjudicating Officers. (c) Rural Electrification Policy with the states and other stake holder finalize in the committee of Secretaries in 2006. (d) 7 Ultra Mega Power Projects With capacity of 4000MW have been evolved and implemented in the capacity addition programme to actuate capacities. (e) National Energy Labelling scheme launched in May 2006.

2007: (a) Electricity (Amendment) Bill, 2007 passed for making electricity theft a cognizable offence and to setup special courts to prosecute power theft and reduce power pilferage, cross-subsidy in power sector slowly and access of electricity to all villages in India. (b) Indian Boilers Act 1923, being amended for inspection of boilers to be carried out during manufacture, erection, operation and maintenance to bring uniformity standards in inspection. (c) States are directed to avail loan under APDRP programme to reduce AT & C losses to 15% or below during 11th plan.

2008: (a) Tax holiday for UMPP's from 2010 to 2017 to sustain interest burden of projects is proposed. (b) Union cabinet's approval for modifications in hydro power policy, and CEA's efforts for efficient design and development. (c) CERC's unscheduled interchange UI ceiling rate from Rs 7.45 per unit to Rs. 10 per unit effective from January 7, 2008. (d) Union cabinets approval for the continuation of RGGVY for eleventh plan period for the development of villages and house hold connections in 2009 and the remaining villages and house holds would be connected by 2012.

RENOVATION & MODERNIZATION (R&M) PROGRAMME FOR THERMAL POWER STATIONS

Introduction of Renovation and Modernization

According to Ministry of Power, in their 2001 Report, circulated, in order to bridge the gap between demand and supply, especially in the context of limited financial resources available, it has become imperative to look for the other options which are not as capital intensive as new capacity addition and which could be implemented in a comparatively shorter time frame. In this regard, optimum utilization of existing installed capacity in the country to maximize the generation, through Renovation & Modernization (R&M) of existing power plants in considered being the most cost effective option.

The report emphasis importance of R&M was recognized by the Government of India way back in 1984 when Phase-I R&M Programme for 34 thermal power stations in the country was launched by the Central Electricity Authority as a Centrally Sponsored Scheme. The Government of India has now accorded high priority to R&M and LE of old existing power plants. The funds, in the form of loan assistance at confessional interest rates, are being provided by the Government through Power Finance Corporation (PFC) for R&M works.

HISTORICAL BACKGROUND OF R&M OF THERMAL POWER STATIONS

Many new units were installed after Independence. Up to 1980, the total installed capacity of thermal power stations was 14,250MW which comprised many imported units of sizes from 30 MW to 120 MW. The average Plant Load Factor (PLF) of thermal power stations during 1976-77 was 56%.

Table 2
Plant Load Factor (PLF) Year Wise

Year	*PLF*
1976-77	55.9%
1977-78	51.4%
1978-79	48.3%
1979-80	44.7%
1980-81	44.6%

Source: CEA & MOP 2001.

However, due to various reasons, the PLF gradually started deteriorating as seen from the table given above. As a result, thermal generation fell short of the targets, leading to a power shortage of about 11%, as compared to the requirement at that time.

ROVING TEAMS OF THE CENTRAL ELECTRICITY AUTHORITY (CEA)

Ministry of Power formed a Task Forces/Roving Teams consisting of experts from CEA, BHEL and ILK were constituted, which went around various thermal stations with a view to identify the defects/reasons for poor performance and to prepare action plans to address the problems. Based on the reports of Roving Teams, many State Electricity Boards (SEBs) prepared schemes for attending to the problems identified and included them in their annual plans. However, lack of adequate funds with some of the States led to lower priority being given to such programmes though there was an urgent need to attend to the problems of the thermal stations.

CAUSES FOR POOR PERFORMANCE

Reasons identified for poor performance of thermal stations in the report include the following:

(i) Design deficiencies, manufacturing and generic defects.
(ii) The Operation and Maintenance (O&M) deficiencies causing prolonged and repetitive forced outages.
(iii) Inadequate and non-timely availability of spare parts especially for imported equipment.
(iv) Lack of resources with SEBs even for making payments to BHEL against supplies and services and for coal supplies to coal companies. Accordingly, they were not able to take up the renovation/modernization programme to the extent required.
(v) The quality of coal being supplied had deteriorated as compared to the designed quality. Besides, the coal had high ash content and contained stones, boulders, shale and sand.
(vi) There was excessive and inadequately trained manpower for he O&M of the plant.

STABILIZATION OF 200/210 MW UNITS

As a result, the performance of these units substantially improved and got stabilized. Because of these efforts, 200/210 MW units are now operating at a level comparable to that of the best in the world. At present, there are 153 units of 200/210 MW comprising 44% of the total thermal installed capacity in the company.

PHASE-I R&M PROGRAMME (7TH PLAN PROGRAMME)

At most of the States were not in a position to arrange adequate funds for R&M works, the CEA formulated a proposal called Phase-I R&M programme as a Centrally Sponsored Scheme for R&M of thermal power stations in the country. Accordingly, the Government of Indian approved the scheme and sanctioned a Central Loan Assistance (CLA) of Rs. 500 crore to provide financial assistance to the SEBs for implementing the core R&M works/activities which could result in direct benefits in terms of generation. The funds for other activities were to be arranged by the states from their own resources. The R&M works relating to environmental improvement in the power plants were also accorded high priority and about 47% of the total investment was made on environmental pollution control measures. The 34 nos. of thermal power stations in various states were covered under the Phase–I R&M Programme. The programme was successfully completed during the 7th Plan and much more benefits than anticipated were achieved.

THE MAIN FEATURES OF THE PROGRAMME

(i)	Number of thermal power stations covered	34
(ii)	Number of thermal units covered	163
(iii)	Total capacity involved 13,570MW	
(iv)	Average PLF of 163 units before R&M Programme	46%
(v)	Anticipated PLF after completion	53%
(vi)	Anticipated additional generation after Completion of R&M Programme 7,000 MU/annum	
(vii)	Year of completion	1990-91
(viii)	Total expenditure incurred Rs.1066 crore (a) Central Loan Assistance Rs.402 crore (b) State Plan Resources Rs.664 crore	
(ix)	Expenditure incurred on environmental activities	47%
(x)	Actual additional generation achieved after completion	10000 MU/annum
(xi)	PLF achieved after completion	56%

Source: MOP

PHASE-II R&M PROGRAMME (8TH PLAN PROGRAMME)

Encouraged with the results of implementation of Phase-II R&M programme, the Phase-II R&M programme for 44 thermal power stations in the country was taken up in the year 1990-91 for completion during the 8th plan period. The CEA prepared the guidelines for preparation of R&M schemes by SEBs and the Roving Teams constituted for Phase-II of the programme visited various thermal power stations and assisted SEBs in the formation of R&M schemes. Power Finance Corporation (PFC) was assigned the responsibility to provide loan assistance to the SEBs for R&M works.

The Progress of work during the initial period remained rather slow, because of paucity, of funds with most of the SEBs, due to their poor financial health. Moreover, they were not initially eligible to avail PFC loans due to its various conditions. However, many SEBs later on became eligible to avail PHC loans after PFC relaxed the conditions. During the 8th plan i.e. by March 1997, about 50% of the work could be completed on all India basis. Some of the balance work was taken up during the 9th Plan and the implementation of some work, which could be postponed for inclusion in the LE schemes planned at a later date, were deferred.

Table 3
Salient Features of the Programme-1

Particulars	*Units*
Number of thermal power station covered	44
Number of thermal units covered	198
Total capacity involved	20870 MW
Estimated Cost	Rs. 2383 crore
Anticipated additional generation after completion	7864 MU/annum
Total expenditure incurred upto March '97	Rs. 862 crore
Total additional generation achieved after partial completion (50%) of work.	5000 MU/annum
Capacity taken up for Life Extension Work (Units: 1-4 of Neyveli TPS)	300 MW

Source: MOP 2001.

R&M PROGRAMME DURING 9TH PLAN

Some of the pending R&M work of the 8th Plan and the additional work subsequently identified were covered under the 9th plan programme. This R&M work on 130 units was at various stages of implementation.

The economic designed life of the thermal power units is considered to be 25 years. 28 nos. of thermal units which had already completed their designed life were based on Residual Life Assessment (RLA) studied during the 9th Plan. The LE work on 19 units has been completed during the 9th plan and work on remaining units is under progress.

Table 4
Salient Features of the Programme-2

Particulars	*R&M*	*LEP*
Number of thermal power stations covered	30	7
Number of thermal units covered	130	28
Estimated Cost	17935 MW	1910MW
Total capacity involved	–	1981 MW
Average PLF of the units before R&M/LE Programme	60.5%	48%
PLF anticipated after completion	62.0%	75%
Anticipated additional generation/annum	2350 MU	4500 MU
Total expenditure incurred	Rs. 651 crore	Rs. 1028 crore
Unit on which LE works completed	–	19 (1343 MW)

Source: MOP 2001.

NATIONAL PERSPECTIVE PLAN FOR R&M AND LIFE EXTENSION OF THERMAL POWER STATIONS

As a long term planning, the CEA formulated a National Perspective Plan (up to 2012) for R&M and Life extension of thermal power station wherein the thermal units required to be taken up for R&M/LE during the 9th, 10th and 11th plans were identified and a total investment of about Rs a. 22000 crore was estimated for the same.

R&M PROGRAMME DURING 10TH PLAN

The CEA has identified 106 thermal of SEBs/State Generation Corporations for LE during the 10th plan so as to extend their economical life by another 15-20 years and to recapture/uprate their capacities by implementing comprehensive R&M Schemes based on RLA studies. Out of 106 units, RLA studied on 46 units have already been completed and further action is being taken by concerned SEBs/Utilities. In addition, another 35 units (6440 MW) which are comparatively new will also need some R&M work to sustain/improve their performance.

Table 5
Salient Features of the Programme-3

Particulars	*R&M*	*LEP*
Number of thermal power stations	12	32
Number of thermal units	35	106
Estimated Cost	Rs.750 crore	Rs. 9200crore
Total capacity involved	6440 MW	10413MW
Average PLF before R &M/LE programme	58.5%	49%
Anticipated PLF after completion	63.5%	75%
Anticipated additional generation/annum	2800MU	23700MU
Capacity expected after completion of LE work.	–	10747 MW

Source: MOP 2001.

R&M PROGRAMME DURING 11TH PLAN

During the 11th plan, 34 Units with a total capacity of 6000 MW are expected to be taken up for LE. Another 31 units with a capacity of 7090 MW would be taken up for R&M Programme to sustain their performance.

Table 6
Salient Features of the Programme-4

Particulars	*R&M*	*LEP*
Number of thermal power stations	12	18
Number of thermal units	31	34
Estimated Cost	Rs. 751crore	Rs. 6000crore
Total capacity involved	7090 MW	6000MW

Source: MOP 2001.

IMPLEMENTATION OF R&M PROGRAMME

(1) Comprehensive R&M/Life Extension Schemes are to be formulated based on RLA studies covering plant performance tests, non-destructive/destructive tests and engineering studies. Latest need-based technologies and improvements for more efficient and reliable operations may be introduced. The environment related activities will be essentially included in the scheme.

(2) The benefits in terms of generation and improved technical parameters anticipated to be achieved after R&M/LE work needs to be identified and incorporated in the bid documents. The schemes should be techno-economically viable.

(3) The RLA studies may be got done through reputed agencies.

(4) As far as possible, LE work where improved technical parameters are to be achieved and demonstrated by the bidder after completion of the work, contract for main plant needs to be awarded as a single point responsibility so as to achieve intended benefits.

(5) Normal procedures of competitive bidding need to be adopted to ensure transparency in process and reasonability of prices.

(6) The CEA is examining the R&M/Life Ectension Schemes for their comprehensiveness, anticipated benefits and techno–economic viability.

(7) The SEBs/Power Utilities, if required, may appoint reputed consultants having engineering background, who could advise them at various stages of the implementation of the R&M/LE Schemes.

(8) The services of NTPC could also be availed by SEBs/ Utilities in the field of O&M of the plants and in R&M work.

(9) The Government of Indian is providing loan assistance through Power Finance Corporation (PFC) at confessional

interest rates for R&M work under AG&SP (Accelerated Generation & Supply Programme)

(10) R&M is not a substitute for regular Operation and Maintenance (O&M) of the plant. There is a need to improve the O&&M practices and carry out regular O&M as per schedules to sustain the improvement obtained through R&M. It should be ensured by the SEBs/Power Utilities that adequate and timely O&M funds are made available to the power stations.

(11) In order to improve the overall work culture, modern management systems including ISO 9002 certification need to be adopted by the power stations. The man behind the machine plays a very significant role in the overall performance of the stations. Therefore, regular training to the power station personnel in the field of O&M and management practices has to be imparted. The 10 th plan plan had achieved 49 to 75% PLF increase though R&M and LE. An additional generation resulted in 23,700 Million Units.

In the eleventh plan period 65 thermal units of 7,090 MW (R&M, LE), in hydro 62 units of 11,340 MW for R&M & 4,392 MW in LE are planned.

ENERGY CONSERVATION

Introduction

Ministry of Power (MOP) in their agenda to prioritize energy as an important element of the infrastructure sector its availability has to be ensured on sustainable basis. On the other hand, the demand for energy is growing manifold and the energy sources are becoming scarce and costlier. Among the various strategies to be evolved for meeting energy demand, efficient use of energy and its conservation emerges out to be the least cost option in any given strategies, apart from being environmentally benign.

(a) Efficient use of electricity and its conservation is the least-cost option to augment the gap between demand and supply of electricity.
(b) Policy goals have to be shifted from "energy' conservation" to "energy efficiency" and from "energy inputs" to the "effectiveness of energy use and energy services.
(c) Improved energy efficiency is the result of many decisions, including those of individual consumers as well as manufacturers and government officials. Gains in energy efficiency in the developed countries in the last three decades were built on a combination of technological improvements, better management practices,
(d) Learning to put these technologies and practices to their best use. In many areas the results have been quite impressive.

Importance of Energy Conservation

MOP tries to accelerate its development process and cope with increasing energy demands, conservation and energy efficiency measures are to play a central role in our energy policy. A national movement for energy conservation can significantly reduce the need for fresh investment in energy supply systems in coming years. It is imperative that all-out efforts are made to realize this potential. Energy conservation is an objective to which I all the citizen in the country can contribute. Whether a household or a factory, a small shop or a large commercial building, a farmer or an office worker, every user and producer of energy can and must make this effort for his own benefit, as well as that of the nation.

Energy Conservation Act, 2001

Recognizing the fact that efficient use of energy and its conservation is the least-cost option to mitigate the gap between demand and supply, Government of India has enacted the Energy Conservation Act-2001 and established Bureau of Energy Efficiency under administrative control of Ministry of Power. The mission of BEE is to develop policy and strategies with a thrust on self-regulation and market principles, within the overall framework of the EC Act with the primary objective of reducing energy intensity of the Indian economy.

Bureau of Energy Efficiency

Bureau of Energy Efficiency have initiated various actions on the thrust areas and moved forward, in establishing institutional mechanisms and infrastructure facilities under the mandate of the EC Act and took steps to involve industries, equipment manufacturers, financial institutions and other stakeholders for implementing the provisions of the EC Act. various activities undertaken by the Bureau of Energy Efficiency have initiated energy conservation in India.

The major user section manufacturers have taken lot of initiatives in energy consumption devises. These are load optimization and demand side management in distribution utilities that help save energy, reduce specific energy consumption in manufacturing and at industries.

ELECTRICITY ACT 2003

Introduction

The first legislation on electricity enacted in 1887 for the protection of a person and properties from any risk or injury over the supply of electricity. This Act are repealed in 1903. An amendment to the Act comes in 1910 with basic legal frame work. 1948 enactment of the Electricity (Supply) Act, dealing with statutory powers and functions of State Electricity Boards (SEB's) and Central Electricity Authority (CEA) and the generating companies. In 1998 Electricity Regulatory Commission Act, enacted to rationalize electricity tariffs for power generation and consumers, to bring in transparency in policy, promote efficient and friendly environment, and to involve private sector in generation and supply. In June 10 the new act enacted as "Indian Electricity Act, 2003", aimed at ushering in competition and efficiency to usher second generation reforms. It consolidates laws on generation, transmission, distribution, trading and use of electricity. It repeal earlier three Acts of 1910, 1948 & 1998 and overrule all previous acts, thus giving clarity to regulators and judiciary. It creates more competitive market and removing many barriers with social interests. The Act is backed up by two major policies-National Electricity Policy and Tariff policy which are being announce by the central government.

Features of the Act

It is mandatory to establish state *regulatory commissions* SERC and central regulatory commission CERC with in six months from June 2003. If required joint regulatory commissions can be established for two or more states if involved. The functions of SERC/CERC include specifying grid code , supply code levy fees and trading margin in interested trading.

Metering of all electricity has been made mandatory within two years of a date notified by Government of India (GOI).

It is mandatory *unbundling* of SEB's into separate corporate entities for generation, transmission & distribution.

Generation has been liberalized and de-licensed. The requirement of techno-economic clearance has been removed except in the case of hydro projects. *Captive plants* have been freed from controls. It has become easy to set up captive power plants and the definition of captive has been enlarged. Consumers also can form "Bulk purchasing groups" and buy power directly from generating companies or from intermediaries such as traders and distribution companies.

The act mandated for non-discriminatory *open access* for transmission and distribution. The CERC duly acted on this and passed orders allowing open access at the national level. Now the states must introduce open access in their respective states in a phased manner. They were given a year's time to decide when and how to introduce open access. States like Kerala allowed open access to its bulk consumer to buy cheap power from power trading corporation as the Indian Aluminum Industry wished to buy. Government has distanced from determination of tariff and tariff fixation has been vested in the CERC/SERC guided by national Electricity policy and Tariff policy. Power Purchase Agreements (PPA's) entered earlier have not been explicitly saved or granted a protection from regulatory intervention.

Distribution has protected *consumer interests* through a consumer redressal forum (CRF) in a time bound manner and can approach ombudsman under SERC. It allows for multiple *distribution* licenses in one circle. The state commissions could grant licenses to two or more entities for distribution of electricity through their own distribution system. The distribution licensee will be free to take up generation and generation companies will be free to take up distribution. Trading has been permitted as a distinct activity. The bill also allows local bodies to like NGO's, punchayats and user associations to undertake distribution of electricity in non-urban areas. A number of private players have shown interest in supply in power in select circles.

In *Rural Electrification* Generation/Distribution require no license in rural notified areas by state governments. Central government shell formulates a national policy in consultation with state Government & CEA.

Trading has bee recognized as a distinct activity. Earlier a transmission licensee was barred from trading. Now the trading of electricity requires a license from Central Electricity Regulatory Authority.

In *transmission* it has been segregated as a wire function without any trading to secure non-discriminatory access. State and Central utilities on transmission are deemed licensees. For efficient, economical and integrated functions National Load Dispatch Centre (NLDC) to coordinate and secure optimum scheduling and dispatch among Regional Load Dispatch Centre (RLDC) and

State Load Dispatch Centre (SLDC) shell integrate operation of power system in state.

The act mandates *Multi Year Tariff* (MYT) framework to be implemented by state utilities on the principles that would result in elimination of cross subsides. This requires detailed cost of service estimate and availability of data on consumption, system peaks, technical and commercial losses to implement a good (MYT) framework unquestionable under law.

Anti Theft Provisions: Power theft is the single largest issue that spoils the power sector. The current poor financial statuses of power utilities are the loss accumulation and the causes of political influence and connivance of Utility Employees with consumers to extract personal benefits at the expense of Utilities. Effective implementation of anti theft provisions are critical for the success of reforms process and being resisted by rested interest.

Generation *tariff*, transmission tariff have been formulated based on well conceived and historical costs that takes care of capital, depreciation, operation and maintenance expenses, return on investment, plant load factor, fixed and variable charges, use of network, System operation, reactive power and surcharge for open access. Instead of government tariff fixation regulatory bodies involve in tariff fixation considering all stake holders.

The Act talks of an *appellate tribunal* to hear complaints against the orders of central and state commissions and exercise supervision and control over the central and state commissions.

Wheeling charges for open access to consumers are computed taking into account projected unit sold and wheeled (Input End Quantity) through distribution licensee network in the ensuing tariff period. The normative distribution system loss to be borne by the consumer in case of wheeling transaction would be based on the voltage levels at which transactions are undertaken. The consumer availing open access would pay surcharge in addition to wheeling charges till cross subsides are eliminated.

The Act gives direction for an efficient, economical and competitive sector under regulatory regime for significant economic development and energy security of India and focuses on the consumer and national economy.

ACCELERATED POWER DEVELOPMENT & REFORMS PROGRAMME

In the process of power sector reforms, government of India initiated a motivational programme to induce unhealthy and sick power state utilities (SEB's), particularly the distribution sector to improve its technical infrastructure in a distribution circle to a profitable circle through assistance financially from a central fund. This programme named Accelerated Power Development & Reforms Programme-(APDRP). Renowned technical expert S. K. Patnaik made a study on the programme.

India's GDP is growing fast. Very strong and sustainable power sector only can target GDP growth of 8% of the country in the coming years. State utilities are suffering losses primarily on account of high technical and commercial losses in the distribution segment. This segment, are the link with other sectors of the economy, this needs support to improve the financial health of the power sector as a whole. A healthy and vibrant distribution segment is the need of the day. Reduction of AT&C losses is the vital issue to improve the financial viability of SEBs/ Discoms, which in turn will provide the payment security to central CPSUs and ultimately pave ways and attract private investment in power sector at internationally competitive tariffs.

The study report explain that, among various distribution reforms initiatives, Accelerated Power Development Programme (APDP) was undertaken from the year 2000-01 for restoring the commercial viability of the distribution segment. It was subsequently rechristened as Accelerated Power Development & Reforms Programme (APDRP) during 2002-03 to provide access to the states to APDRP funds on the basis of agreed reforms program to narrow down losses.

While the Electricity Act 2003 paves way for open access for deriving economic benefit from available captive distributed generation available in the country to arrive at competitive tariffs, APDRP aims at turnaround of distribution utilities, with the ultimate objective of customer satisfaction are noted by the study.

Accelerated Power Development and Reforms Programme (APDRP)

The APDRP was operationalized by the Govt. of India with the following objectives as narrated by the author are:

(i) Reduction of AT&C losses

(ii) Bringing commercial viability

(iii) Reduction of outage and interruption

(iv) Improved customer satisfaction

In this funds are provided through state governments to the SEBs/Discoms which have adopted the path of Distribution reforms for up-gradation and strengthening of the Sub-Transmission and Distribution (ST&D) system so as to achieve the objectives of APDRP.

The study narrates, Funds are also provided as 100% grant under incentive component to the states as additional plan assistance. The state governments offer incentives up to 50% of the actual total loss reduction by the SEBs/utilities, as grant. The year 2000-01 is taken as base year for calculation of loss reduction. Losses are calculated net of subsidy, if any, given by the state government and revenue is considered on net realization basis.

During Xth plan, an outlay of Rs. 40,000 crores as additional Central Plan assistance to SEBs/Discoms through State Government was planned, which was further divided as:

(a) Rs. 20,000 crores for strengthening and up gradation of Sub-transmission and Distribution Network under investment component to reduce AT&C losses and improve reliability of network and improved customer satisfaction. Though initially schemes were circle based, subsequently the focus shifted to urban towns for quantification and showcasing results. Since FY 2006-07, GoI provides central plan assistance as 25% of project cost as grant and remaining 75% of project cost to be arranged by utility as loan from FIs like PFC, REC, etc. Earlier, 50% of the project cost was provided by GoI as grant (25%) and loan (25%), and the remaining counterpart fund was to be arranged by the utilities. Special category states (All North Eastern states, J&K, H.P., Uttaranchal and Sikkim) are entitled for 90% of project cost as assistance in the form of grant and have to arrange balance 10% from internal sources/FIs.

(b) Rs. 20,000 crores grant to the states for reduction in the cash loss by SEBs/Utilities under incentive component to motivate utilities to reduce cash losses through technical/administrative/commercial measures in order to encourage utility to move towards financial viability. Total incentive release by MoP to states of AP, Maharashtra, Gujarat, Rajasthan, Haryana, West Bengal, Punjab and Kerala amounted to Rs. 1536 crores till financial year (FY) 2005-06. In XIth plan it is directed to make 15% AT & C loss.

Memorandum of Agreement (MoA)

It put it that, utilities willing to reform are entitled to APDRP funds and has to enter into Memorandum of Agreement with GoI for taking up Technical, Commercial and Administrative measures in a time bound manner to achieve financial turnaround with improved customer satisfaction. Primarily, the ingredients of a MoA are as under.

ADVISOR CUM CONSULTANTS

As per the study, in 2001, Ministry of Power, GoI appointed NTPC and POWERGRID as the lead Advisor cum Consultants (AcC) and subsequently, brought CPR I and MECON under the charge of NTPC to bring in uniformity and consistency in approach amongst AcCs for 12 States (AP, Kerala, Karanataka, TN, MP, Maharashtra, Punjab, Uttaranchal, Haryana, Rajasthan, Jharkhand and Chattisgarh) with the following responsibilities:

- Advise and help SEBs/utilities in preparation of DPRs.
- Assist SEBs/Utilities in prioritization of schemes for urban areas in distribution circles and ranking of projects for improvement in revenue collection and increase in customer satisfaction.
- Assist the Monitoring Committee for review of techno economic appraisal of the DPRs prepared by utilities.
- Assist MoP in prioritization of available funds to be allocated to States/Utilities under APDRP scheme.
- Monitor periodically the physical and financial progress of various distribution circles for which APDRP funds have been allocated and disbursed.
- Advice MoP in devising guidelines for uniformity of approach in preparation and evaluation of project proposals, and disseminating information on best practices under APDRP scheme.
- From financial year 2002-03, NTPC has taken the overall responsibility of implementation of Gol's APDRP scheme under the guidance of MoP.

NTPC to represent in Distribution Reforms Committee of states consists of members from State Government, Utilities and NTPC on behalf of MoP. Distribution Reforms Committee monitors implementation of schemes, reforms initiatives taken by the States.

The study concludes that the programme, calls for enthusiasm of field staff and commitment of top management in implementation of policies and guidelines besides bringing in reforms in distribution business process management for achieving transparency and efficiency. Speedier projects execution, administrative initiatives under Electricity Act 2003 and IT enabling business processes for reduction of human intervention are ushering in enabling environment to turnaround Distribution Utilities.

Study explains that to impose culture of accountability and responsibility in staff, energy audit at DT level through time-stamped energy metering system with respect to GIS based indexed consumer and network data is the imperative need of the hour to identify loss pockets to determine baseline benchmark performance parameters and monitoring improvement of the same. APRDP is instrumental in turnarounds and substantial improvements in financial health of a number of utilities across the country. The study explain APDRP programme to sustain full benefits of power sector in long run.

CONCLUSION

To sum up primary financial disorder in power sector is long term tariff that are below actual cost of power supply. In the initial phase of power reform process only 80% recovery on tariff was possible. Cross subsidy have been targeted to gain vote banks from domestic and agriculture consumers by politicians. Average cost of supply is 3.50 Rupees/KWh and average tariff is only 2.40 Rupees/KWh (Planning Commission 2002) leading to a huge SEB loss that are now over 1% of GDP. Parts of these losses on under charged tariff are provided from state budgets. Total arrears to state owned enterprises from SEB's on power purchase from generators are 340 billion (Ahlluwalia 2001). When SEB's sought expansion through internal accruals, debt or equity they ran dry of financials. Only government grant and debts were possible routes for expansion. At this stage power investments by state sector are in logger heads. Central government plans for reformation on distribution that set right revenues and cost of power for a good recovery to boast the power sector of good financial status through distribution sector modifications that are being assisted by funding through central agencies.

According to MOP sources, as of January 2008, Rs. 7,124.63 crores has been released to the states so far under APDP. Under

APDRP 5.49 billion and 3.84 billions were released as investment and incentive respectively in 2007. In 2007 till December, 39,383 villages electrified and over 1.8 million rural house hold were electrified under RGVY and franchisees were also deployed for this purposes. The sector is heading towards a long leap for future power demands. This is a complex situation. Reformation has shown path and lead to a healthiness. The passage is way ahead.

3

PERFORMANCES OF THERMAL POWER PLANTS INDIA: OVERALL POWER GENERATIONS

Generation of power is indicated by Kilo Watt per hour (Kwh) or one unit. These are also indicated in million units for the bulk quantities.

The Overall generation (or total) in the country has increased from 5.1 Billion units (BU) in 1950 to 617.4 BU in 2005-06. Overall power generation include all sources of power generations form thermal, hydro, nuclear, renewables.

The overall power generation are made in different sources of power such as thermal, hydro, nuclear, renewable sources. Pre reforms and post reform generation are before the year 1990 and beyond this period s respectively.

The generation from all sectors thermal, hydel, nuclear and other source has a growth from 5.1 billion units in 1950 to 617.4 billion units in 2005-06. There is a marginal raise from 5.1 to 240.34 billion units in 40 years from 1950 to 1990 in pre reforms period and 264.3 to 617.4 billion units steep raise in post reform period.

The steep raise in post reforms period in overall power generation attributes to the better performance measures to increase power generations thorough various means and methods of generations adopted over the years at the initiative steps of CEA and Ministry of Power. The over all power generation are analyzed above using twenty two years generation in billion units showing a linear trend of the following type.

Table 1
Overall Power Generations since 1950–2006
The overall power generations in year wise are given below

Year	*Generation (Billion Units)*	*Variation*	*Percentage*
Pre reforms period			
1950	5.1		
1955-56	9.66	4.56	89.41
1960-61	16.94	7.28	75.36
1965-66	32.99	16.05	94.74
1968-69	47.43	14.44	43.77
1970-71	55.83	8.40	17.71
1973-74	66.69	10.86	19.45
1975-76	79.23	12.54	18.80
1978-79	102.52	23.29	29.39
1979-80	104.62	2.10	2.04
1980-81	110.84	6.22	5.94
1984-85	156.86	46.02	41.51
1985-86	170.35	13.49	8.60
1989-90	245.44	75.09	44.08
Post reforms period			
1990-91	264.3	18.86	7.68
1995-96	380.1	115.8	43.81
2000-01	499.5	119.4	31.41
2001-02	515.3	15.8	3.16
2002-03	531.4	16.1	3.12
2003-04	558.3	26.9	5.06
2004-05	587.4	29.1	5.21
2005-06	617.4	30.00	5.10

Source: MOP & CEA 2006.

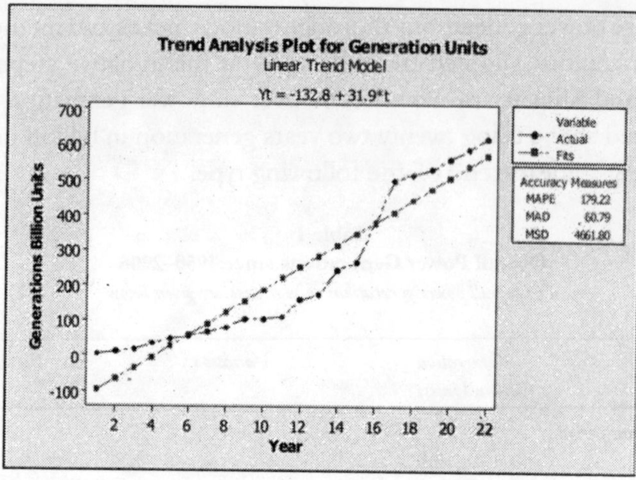

Figure 1 : *Overall Power Generations since 1950–2006*

Thermal Power Generations from 1950 to 2006

Thermal generation capacity from thermal power generation alone, accounts for a very high 81% of the total generations. Year wise generations of thermal power are given above, in million units.

Thermal power generations are analyzed using twenty five years thermal power generations in million units showing a linear trend of the following type. Thermal power generations before reforms, as shown in, 1970-71 to 1889-90 have increased year wise, the raise in thermal generations over the years increased from 28160 million units to 178696 million units, over 19 years it show an average of 9405 million units per year. Thermal power generations after reforms, beyond 1990 have increased year wise, the raise in thermal generations over the years increased from 186550 million units to 500006 million units, over o period of 16 years show an average generations of 31250 million units per year. Pre reforms period slow growth attribute to thermal plants low PLF, lower operational performances and lower capacities available. High variations shown in 1970-71 to 1974-75, 1981-85, and 1985-86 to 1989-90 are due to five years period variations. Post reforms higher growth causes are more of capacities, higher PLF, higher operating efficiencies. In 1995-96 see higher generation are due to more of capacity additions. Other wise only average accelerated growth of generations will takes place.

Table 2

Year	*Generations (MU)*	*Variation*	*Percentage*
1970-71	28160		
1973-74	35321	7161	25.42
1975-76	43300	7979	22.58
1978-79	52594	9294	21.46
1979-80	56273	3679	6.99
1980-81	61300	5027	8.93
1984-85	98936	37636	61.39
1985-86	114350	15414	15.58
1989-90	178696	64346	56.27
Post reforms period			
1990-91	186550	7854	4.39
1991-92	208747	22197	11.89
1992-93	224680	15933	7.63
1993-94	248100	23420	10.42
1994-95	261940	13840	5.58
1995-96	298820	36880	14.07
1996-97	316162	17342	5.80
1997-98	335110	18978	6.00
1998-99	352585	17475	5.21
1999-00	385326	32741	9.28
2000-01	408359	23033	5.98
2001-02	422001	13642	3.34
2002-03	448450	26449	6.27
2003-04	466824	18374	4.10
2004-05	486075	19251	4.12
2005-06	500006	13931	2.87

Source: MOP & CEA 2006- (MU-Million Units).

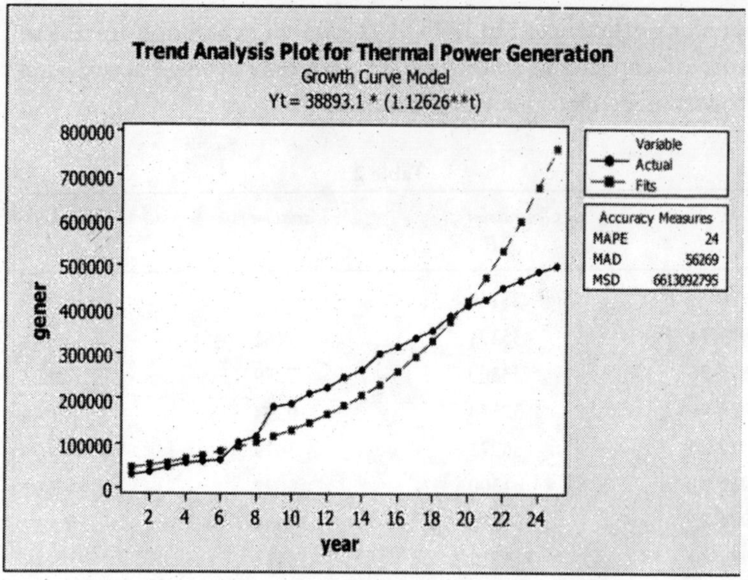

Figure 2 : *Thermal Power Generations from 1950 to 2006*

Overall Power Generations and Thermal power Generations, and Share of Thermal in Total Generations

A comparison of over all generation and only thermal generation are given below.

The slow growth in pre reforms period is 72% in 9 years attribute to low efficiencies, low PLF%, higher outages and availability factors. Apart from this fuel arrangements and shortages also are causes. Compared with overall generations and only thermal generations in the year 2006, they are 81%.

Before reforms and after, the generation percentages are 72% to 81%. This indicates a steep growth in thermal generation after reforms period.

Higher PLF thrust, efficient plant managements, planned production techniques are the high generation causes. They also attributes to capacity increase, plant operation performances and fuel availabilities.

Table 3
Overall Power Generations and Thermal Power Generations (Pre Reforms Period & Post Reforms Period)

Year	*Overall Generation (BU)*	*Thermal Generations (BU)*
Pre reform		
1970-71	55.83	28.16
1973-74	66.69	35.32
1975-76	79.23	43.36
1978-79	102.52	52.59
1979-80	104.62	56.27
1980-81	110.84	61.3
1984-85	156.86	98.93
1985-86	170.35	114.35
1989-90	245.44	178.69
Post reforms		
1990-91	264.3	186.55
1995-96	380.1	298.82
2000-01	499.5	408.36
2001-02	515.3	422
2002-03	531.4	448.45
2003-04	558.3	466.82
2004-05	587.4	486
2005-06	617.4	500

Source: CEA & MOP 2006.

As the post reform give thrust on generations and operating efficiencies they give a push in thermal plants and have steep generation growth.

Table 4
Correlation of Thermal Generations and Overall Generations

		Thermal Generation	*Overall Generation*
Thermal Generation	Pearson Correlation	1	.999(**)
	Sig. (2-tailed)		.000
	N	17	17
Overall Generation	Pearson Correlation	.999(**)	1
	Sig. (2-tailed)	.000	
	N	17	17

** Correlation is significant at the 0.01 level (2-tailed).

There is a very close relation between the thermal generations and overall generations in Indian Power sector. By Correlating the Thermal Generation and Overall generation the 'r' value is 0.999.

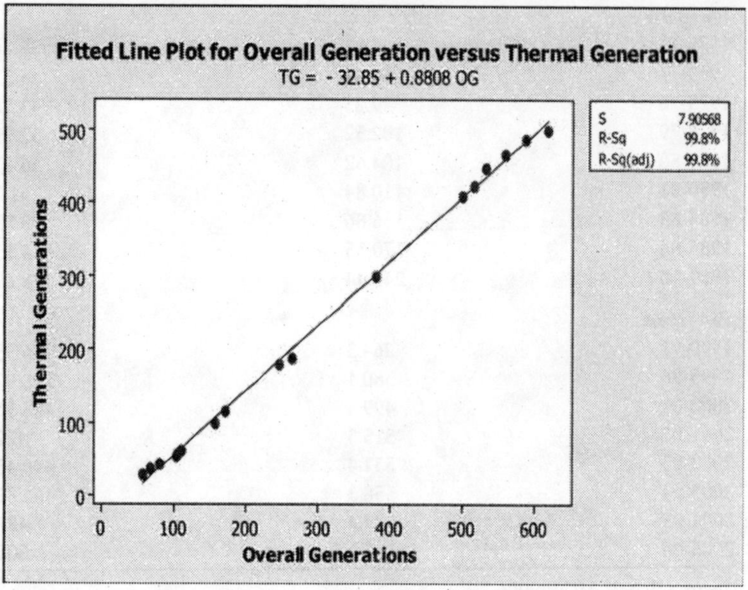

Figure 3

OVERALL GENERATIONS AND THERMAL GENERATION

Correlation of the Thermal Generation and Overall generation the 'r' value is 0 .999.

Plant Load Factors & Detoriations

Plant load factors (PLF) are the performance indicators of a plant. After stabilization period of SEB's their performances start sagging and a detoriation of PLF from 55.9% to 44.60% in 1976 to 81 are given below.

The sag in PLF are 55.9% in 1976-77 to 44.60 % in 1980-81. The sag or low PLF % are indication of lower plant level efficiency in pre reform period. The SEB's working has gone down to a low level. Plants operated at low efficiency levels. Many plants start showing detoriation that has led to plant up-keepment, renovation and modernization of old plants and performance development programme. Government has evolved plans to increase performances. These renovations and modernization programme conducted by expert professionals for efficient working. This has started improving PLF of thermal plants. PLF have moderately increased and are standardized.

Table 5
Plant Load Factors & Detoriations

Year	*PLF %*
1976-77	55.90%
1977-78	51.40%
1978-79	48.30%
1979-80	44.70%
1980-81	44.60%

Source: CEA 1981.

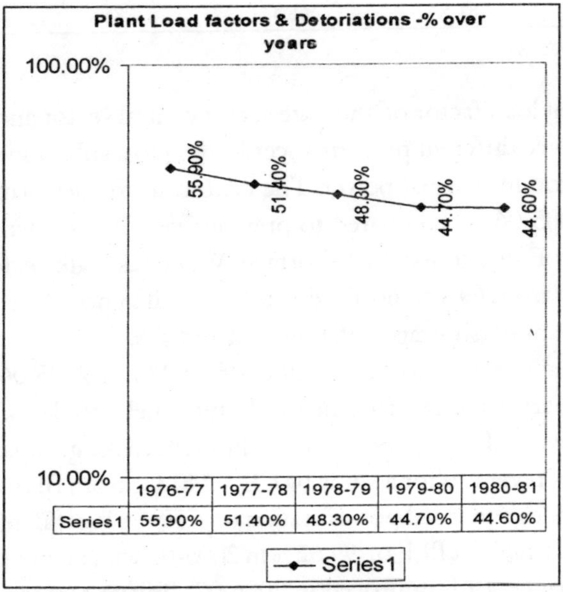

Figure 4 : *Plant Load Factors & Detoriations*

Thermal Utilities PLF Performances Sector Wise

Thermal utilities sector wise PLF% of central sector, State sector, private sector performances are detailed as below:

Table 6
Thermal Utilities PLF Performances Sector wise

Year	*Central*	*State*	*Private*	Overall
1985-86	61.9	49.2	57.5	52.4
1989-90	62.2	53	69.9	56.5
1990-91	58.1	51.3	58.4	53.8
1992-93	62.7	54.1	54.1	57.1
1994-95	69.2	55	65.9	60
1995-96	70.9	58	72.3	63
1996-97	71	60.3	71.2	64.4
1998-99	71.1	60.8	68.3	64.6
2000-01	74.3	65.6	73	69
2001-02	74.3	67	74.7	69.9
2002-03	77.1	68.7	78.9	72.2
2003-04	78.7	68.4	80.4	72.7
2004-05	81.7	69.6	85.1	74.8
2005-06	81.5	66.4	85.1	73.71

Source: CEA 2006.

Plant load factor of the state sector central sector and private sector have different performances leading to a substantial power generation in thermal power. The central sector performed 81.5 % in 2005-06, as compared to previous year had shown slightly lower PLF due to gas fuel shortage. Where as state sector's had declined from 69.4 to 66.4% due to shortfall in power generation on account of shortages of fuel-coal and gas.

Private sector has higher PLF of 85.1% in 2005-06 due to higher performance efficiencies, despite fuel problems. BHEL make have PLF of 76.59% among different make groups. Out of 380 units reviewed for a study by CEA 240 are from BHEL make. Budge Budge thermal power station of CESC , West Bengal have achieved a highest PLF of 99.62% in 2005-06 with an out standing performance. PLF improved in 2006-07. Private sector achieved highest PLF of 86.35 followed by central and state sectors.

Combined Plant Load Factors (as Percentage)

Plant load factors are the performance indicator of a power plant. Plant load factor after a slump in mid 1970' and 1980's has an impact on power generations. Year wise PLF % are given below.

Table 7
Combined Plant Load Factors-as Percentage

Year	*Plant Load Factor (PLF) in %*	*Variations*	*Percentage*
Pre reforms period			
1976-77	55.9		
1977-78	51.4	-4.5	-8.0
1978-79	48.3	-3.1	-6.04
1979-80	44.7	-3.6	-7.45
1980-81	44.6	-0.1	-0.22
1985-86	52.4	7.8	17.49
Post reforms period	-		
1989-90	56.5	4.1	7.82
1990-91	53.8	-3	-5.31
1992-93	57.1	3.3	6.13
1994-95	60	2.9	5.07
1995-96	63	3	5
1996-97	64.4	1.4	2.22
1998-99	64.6	0.2	0.31
2000-01	69	4.4	6.81
2001-02	69.9	0.9	1.30
2002-03	72.2	2.3	3.29
2003-04	72.7	0.5	0.69
2004-05	74.8	2.1	2.89
2005-06	73.71	-1.09	-1.45

Source: CEA 2006.

The PLF % has detoriated from 55.9 in 1976-77 to 44.6 in 1980-81. Operational inefficiencies, frequent shutdowns, breakdowns, technical designs, lack of renovation and modernizations plans, old designs plant maintenance plan are the causes. Various measures of R&M programme have improved the PLF of thermal plants in 1985-86 to 1989-90 to 56.5% after this programmes are implemented.

The reforms have given thrust on plant efficiencies. Various measures have given improved plant atmosphere and efficiencies. These efficiency factors had led to a remarkable result in the form of plant performances. The PLF % has an increase of 73.71 % in 2005-06 from a low 53.8%. in 1990-91. This is important improvement that have increased generation substantially.

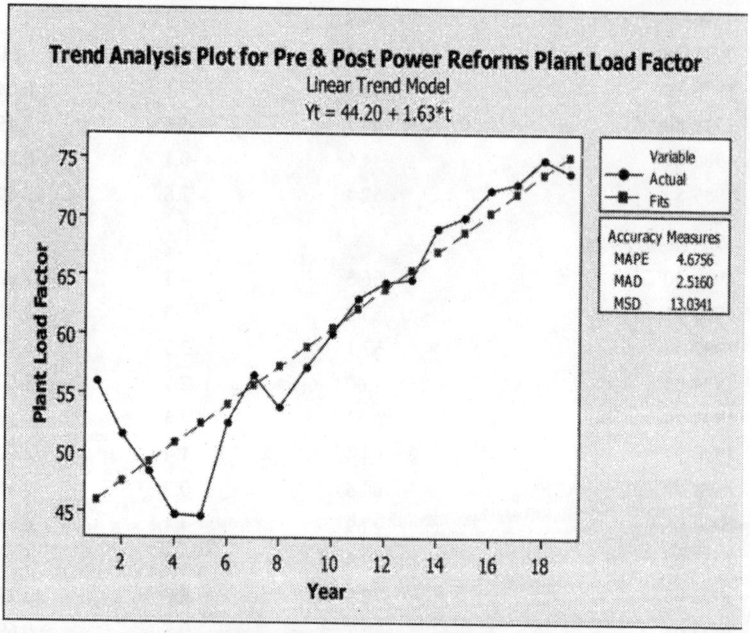

Figure 5 : *Combined Plant Load Factors (as %)*

Generation Capacity Wise Thermal Utilities PLF Performance

Capacity group wie PLF % are given below for 2003-04 to 2005-06. 250 MW capacity groups are having highest PLF of 87.74% among other groups. 500 MW capacity have next highest PLF of 82.73% among the group. 200/210 MW have 79.22 %. Others have low PLF %.

Table 8
Generation Capacity Wise Thermal Utilities PLF Performance

Capacity Group (MW)	*No of Units*	*Capacity (MW)*	*Reviwed Units*	*Capacity (MW)*	*PLF %03-04*	*PLF % 04-05*	*PLF % 05-06*
500	33	16500	31	15500	82.36	84.32	82.73
250	13	3250	13	3250	86.48	90.19	87.74
200/210	159	33145	158	32935	79.44	79.76	79.22
140/150	9	1270	9	1270	42.21	55.76	50.52
120	23	2770	22	2645	34.86	42.89	44.46
110	40	4400	40	4400	53.64	55.92	52.84
100	11	1100	11	1100	52.01	50.21	51.27
70/85	8	630	8	630	49.58	45.62	54.23
62.5/67.5	25	1601.5	25	1601.5	61.79	66.9	59.41
60	22	1320	22	1320	44	46.32	42.27
50/57.5	27	1337.5	27	1377.5	38.99	42.7	35.62
20/40	14	420	14	420	37.63	39.9	44.65
Total	384	67784	380	66449	72.96	74.82	73.71

Source: CEA 2006.

High Performers in 2005-06 in Thermal Power

Countries high performing thermal plants for the year 2006 are given with PLF % achieved. Budge-Budge's thermal plant show outstanding performances, followed by Dhanu thermal are remarkable. Coal and Gas shortages in last two years resulted in top performing plants end in lower performances.

High Performers in 2004-05 in Thermal Power

Country's high performing thermal plants for the year 2004-05 are given with PLF % achieved.

Dahanu (Reliance Energy) is the outstanding thermal power station in the year 2004-05 with more than 100% PLF performances. NTPC's plants stand outstanding in the next grades.

Table 9

Plant	*State*	*Capacity (MW)*	*Generation (Mus)*	*Capacity engaltion* (%)*	PLF (%)
Coal					
Budge Budge	West engal	500	4362.72	0	99.6
Dahanu	Maharashtra	500	4323.11	0	98.7
Unchahar	Uttar Pradesh	840	7041.1	0	95.7
Ghandhinagar-5	Gujarat	210	1743.46	0	94.8
Sabarmathi	Gujarat	330	2703.14	0	93.5
Wanakbori	Gujarat	210	1709.31	0	92.9
Vindhyachal STPS	Madhya Pradesh	2260	18304.6	0	92.5
Dadri NCTPP	Uttar Pradeh	840	6768.3	0	92
Korba-II	Chhattisgarh	200	1610.63	0	91.9
Suratgarh	Rajasthan	1250	9951.25	0	90.9
Kahalgaon	Bihar	840	6572.3	0	89.3
Gas					
GTE Corporation	Gujarat	665	4755.99	82.89	0
Pragati CCGT	Delhi	330	2298.07	79.4	0
Gandhar GT	Gujarat	648	4478.2	78.89	0
Faridabad CCGT	Haryana	430	2952.8	78.39	0
Anta GT	Rajasthan	413	2809.1	77.64	0
Vijjeswaran GT	Andra Pradesh	272	1836.92	77.09	0

Source: CEA 2006

Table 10

High Performers in 2004-05 in Thermal power

Comparative plant performance-10 best performing power stations in India (2004-05)

Station	*Capacity*	*Gen.*	*PLF*	*Avail.*	*Loading factor*	*Aux. Power*	*Specific oil*	*Heat rate*	*Total outages*
Promoter	(MW)	(MUs)	(%)	(%)	(%)	(%)	(ml/kWh)	(kcal/kWh)	(No.)
Dahanu (REL)	500 (2x250)	4,439.2	**101.35***	96.88*	104.61*	7.53	0.139	2,261*	14
Kothagundam New (Ap Genco)	500 (2x250)	4,140	**94.53**	94.74	99.78	8.95	0.430	—	30
Dadri Coal (NTPC)	840	6,831.2	**92.84**	95.77	96.94	7.37	0.160	2,421	24
Simhadri (NTPC)	1,000	8,122.7	**92.72**	93.23	99.45	6.86	0.220	2,391	18
Korba (NTPC)	2,100	17,050	**92.68**	93.38	99.25	8.59	0.110	2,402	26
Unchahar (NTPC)	840	5,781	**92.15**	92.25	99.89	8.56	0.430	2,451	44
Rayalaseema (Ap Genco)	420	3,353	**91.16**	92.97	98.05	9.37	0.226	2,288	11
Rihand (NTPC)	1,000	7,956.7	**91.10**	92.27	98.73	7.97	0.170	2,378	19
Ramagundam (NTPC)	2,100	17,172	**90.51**	91.04	99.42	8.81	0.170	2,424	28
Singrauli (NTPC)	2,000	15,603	**90.20**	91.69	98.37	8.87	0.300	2,412	45

Source: MoP.

AUXILIARY POWER CONSUMPTION

Auxiliary power consumption are the power consumed to operate the power plant auxiliary equipments and lighting system. Year wise auxiliary consumption before and after reform period are given a below.

Table 11
Auxiliary Power Consumption

Year	*Auxiliary Consumption %*	*Variations*	*Percentage %*
Pre reform period			
1980-81	10.4		
1985-86	11.1	0.7	6.73
1989-90	9.9	-1.2	-10.81
Post Reforms period			
1996-97	8.91	-0.99	-10
1997-98	8.86	-0.05	-0.56
1998-99	8.79	-0.07	-0.79
1999-00	8.81	0.02	0.23
2000-01	9.01	0.2	2.27
2001-02	8.72	-0.29	-3.22
2002-03	9.55	0.83	9.5
2003-04	9.91	0.36	3.77
2004-05	8.57	-1.34	-13.52
2005-06	8.44	-0.13	-1.52

Source: Planning Commission 2002.

In 1980 to 90 reduction in auxiliary consumption started in thermal plants. Coal plant's high ash contents, coal handling plant's initial practices and its installations, earlier operation and maintenance practice and technology are the causes.

Earlier operation and periodic maintenance vary with new schedules. The higher consumption on auxiliary and its cause are slowly start analyzed to reduce consumptions. Auxiliary power consumption of thermal plants are in higher side in India as compared to other countries. As the Indian coal has high ash content and as a result of higher consumption in coal handling and fume extraction systems occured. Reduction in partial loading units and avoid running extra auxiliary equipments are required. BHEL/BHEL units have less consumption of 8.30% followed by Russia/Russia 8.40% and BHEL/ABL make have highest 10.22% auxiliary consumption. Northern region have highest 8.89%, North Easte region has lowest 6.42%. Pre and post reforms period auxiliary consumption show a reduction from 11.1% to 8.44%.

The 13 years Auxiliary Consumption in percentage showing a linear trend of the following type.

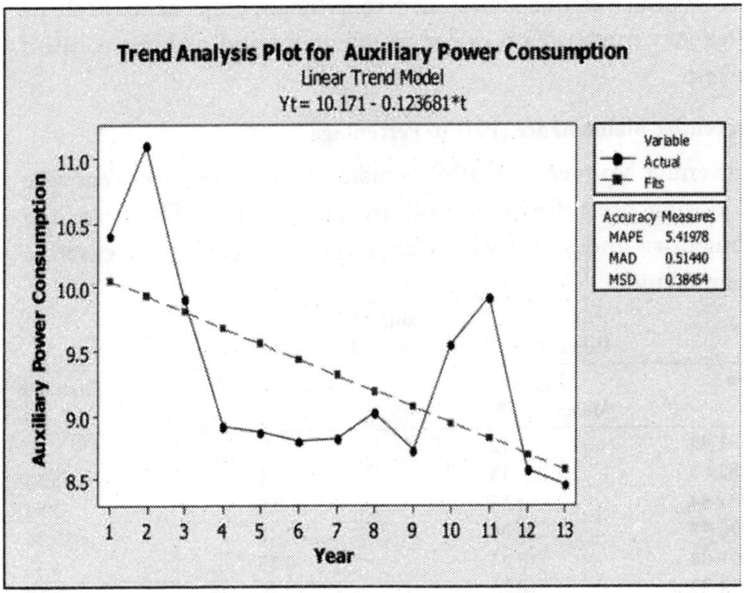

Figure 6 : *Auxiliary Power Consumption*

Preventive/Planned Maintenance (PM): (Sector wise)

It is a performance Indicator for power plant and it is the level of planned maintenance. Sector wise preventive maintenance are given for the years 2000-01 to 2005-06 as below:

Table 12
Preventive/Planned Maintenance (PM): (Sector wise)

Sector	*2000-01*	*2001-02*	*2002-03*	*2003-04*	*2004-05*	*2005-06*
Central	7.18	6.78	8.18	8.35	7.03	6.85
State	11.27	11.52	8.71	9.04	9.49	11.73
Private	3.5	4.88	4.3	4.9	2.18	3.34

Source: CEA.

State sector Preventive maintenance remained highest with 11.73% in the last 6 years. Private sector recorded a low 3.5% in PM since lat 6 years. The central sector's preventive maintenance are at the next stage of middle range of 6.78%, among state, central and private sectors. The lower percentage performance of state and central sector's Preventive maintenance are as a result of efficiency primarily on boiler maintenance and turbine overhaul system.

Preventive Maintenance (PM) in Percentage

Preventive Maintenance (PM), is indicated in percentage. Year wise % during pre reforms period are given below. These are the obligatory and statutory overhaul and preventive or curative maintenance.

Table 13
Preventive Maintenance (PM) in Percentage

Year	*Preventive Maintenance %*	*Variation*	*Percentage*
1981-82	12		
1982-83	13	1	8.33
1983-84	12.5	-0.5	-3.84
1996-97	10.08	-2.42	-19.36
1997-98	10.43	0.35	3.47
1998-99	10.71	0.28	2.68
1999-00	10.21	-0.5	-4.67
2000-01	9.44	-0.77	-7.54
2001-02	9.52	0.08	0.85
2002-03	8.3	-1.22	-12.82
2003-04	8.59	0.29	3.49
2004-05	8.23	-0.36	-4.20
2005-06	9.48	1.25	15.18

Source: Performance Review of thermal stations 2005-06.

Out of 12-13 %- that prevailed in pre reform period, 2% are preventive maintenance and the balances are obligatory or statutory maintenance. Boiler over haul takes 50-60 days and 120-125 for capital maintenance. Reduction plan to reduce this %. Proper inventory and stock of frequently used spares are taken care to make available supply for this maintenance. Schedule of maintenance are started done involving all departments. Pre reforms preventive maintenance % varied at 12 to 13%. These are on higher side. CEA's roving team made special efforts to reduce the pre reform % of planned shutdown. The post reform preventive maintenance % has come down from 13% to 8.23%. PM of 380 units recorded 9.48% as against programmed 11.2%. The average duration of boiler overhaul was 29 days. The average duration of capital maintenance of Turbine was 66 days. More attention for 200/210 MW category are given.

Their boiler overhauls are 101 and turbine overhaul are 15, with average duration of 26 and 49 days respectively. Average duration of boiler overhaul in private sector are lowest-21 days, followed by central sector 22 days and state sector 34 days.

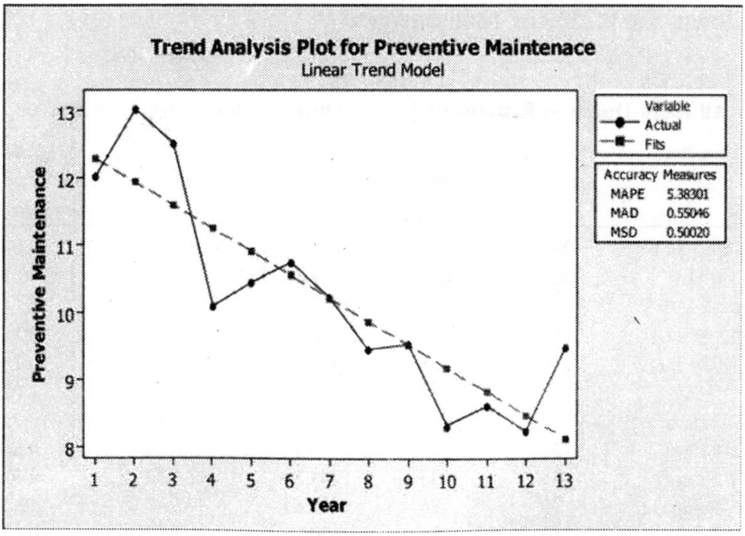

Figure 7: *Preventive Maintenance (PM) %*

Forced Outages (F.O.)

Forced outages are the performance indicators of a power plants. They are unplanned component failure or any other condition that requires generating unit to be removed from services immediately or with in a short period of time.

Table 14
Sector Wise Forced Outages

Sector	*2000-01*	*2001-02*	*2002-03*	*2003-04*	*2004-05*	*2005-06*
Central	9.52	9.63	7.52	6.86	5.88	6.37
State	11.51	10.18	11.18	10.68	10.4	10.75
Private	4.85	4.06	4.28	3.32	4.95	3.57

Source: CEA 2006.

The state sector outages are highest followed by central and private sectors. Central sector are able to reduce 3% in the last five years. Private sector hover around 4%. Boiler and auxiliaries account for major losses in generations. Boiler and auxiliaries accounted for 39.39 % in 2005-06. Generator problems are 6.05% and the share of turbine are 21.62% in 2005-06. Reserve shut downs are 9.21% of total outages.

Table 15
All India-Duration Patterns of Forced Outages During the Year 2005-06

Duration	*No of units*	*No of outages*	*MU Loss*	*% of All India F.O. Loss*	*% of All India Poss. Generation*
Upto 3 Hrs	292	2022	448.38	0.89	0.08
3-6 Hrs	266	839	597.66	1.18	0.1
6-12 Hrs	235	647	800.19	1.58	0.14
12-18 Hrs	159	342	655.38	1.3	0.11
18-24 Hrs	149	288	869.06	1.72	0.15
1-2 days	281	1116	5892.22	11.66	1.02
2-3 days	213	601	5495.97	10.87	0.95
3-4 days	135	240	3255.05	6.44	0.56
4-5 days	85	124	1903.84	3.77	0.33
5-6 days	70	86	1755.64	3.47	0.3
6-7 days	43	51	1003.3	1.99	0.17
7-8 days	32	36	1007.23	1.99	0.17

Contd...

Duration	*No of units*	*No of outages*	*MU Loss*	*% of All India F.O. Loss*	*% of All India Poss. Generation*
8-9 days	21	23	567.05	1.12	0.1
9-10 days	21	23	631.59	1.25	0.11
10-11 days	14	15	451.15	0.89	0.08
11-12 days	21	22	986.67	1.95	0.17
12-13 days	9	9	465.25	0.92	0.08
13-14 days	9	9	483.38	0.96	0.08
14-15 days	13	13	654.72	1.3	0.11
15-16 days	8	8	535.99	1.06	0.09
16-17 days	6	6	263.78	0.52	0.05
17-18 days	67	7	358.17	0.71	0.06
18-19 days	4	4	165.1	0.33	0.03
19-20 days	6	7	355.25	0.7	0.06
20-21 days	5	5	373.24	0.74	0.06
21-22 days	3	3	151.05	0.3	0.03
22-23 days	1	1	58.88	0.12	0.01
23-24 days	4	4	220.34	0.44	0.04
24-25 days	3	3	195.64	0.39	0.03
>25 days	56	67	19941.39	39.45	3.45
Grand Total	354	6621	50543.54	100	8.74

Source: CEA.

Number of units Considered: 380 Number of units involved: 345. All India Possible gen (MU):578099.04 All India F.O.Loss (MU): 50543.54. Effect of forced outages on generation reliability during 2005-06. Month wise Annual and Capital Maintenance schedule could not be adhered to in actual practice and therefore generation reliability was adversely affected.

Table 16

Capacity Group-wise Means

Time Between Failures (MTBF) During 2005-06

No of units involved

MTBF Range (Days	*500*	*250*	*200/ 210*	*140/ 150*	*120*	*110*	*100*	*<100*	*All Cap.*
Above 30	21	4	62	0	1	5	4	14	111
25-30	6	4	14	1	1	1	0	5	32
20-25	1	2	18	1	2	5	1	8	38
15-20	2	1	24	3	4	1	1	8	44
15-Oct	1	0	24	3	9	4	0	11	52
10-May	0	0	12	1	2	5	2	15	37
5-Jan	0	2	4	0	0	13	1	13	33
UPTO 1	0	0	0	0	3	6	2	22	33
Total	31	13	158	9	22	40	11	96	380

Source: CEA 2006.

The mean time between failures of various capacity are given as all capacities total failures as 380 days.

Thermal Plants Forced Outages

Forced outages are caused due to removal of equipment deficiency that accounted for 80% of the unscheduled outages during 1980's period.

Table 17
Thermal Plants Forced Outages

Year	*Forced Outages %*	*Variation*	*Percentage*
1981-82	20		
1982-83	19	-1	-5
1983-84	16.5	-2.5	-13.15
Post reform period			
1996-97	10.79	-5.71	-34.60
1997-98	10.79	0	0
1998-99	10.41	-0.38	-3.52
1999-00	9.9	-0.51	-4.9
2000-01	10.47	0.57	5.76
2001-02	10.56	0.09	0.85
2002-03	9.87	-0.69	-6.53
2003-04	9.48	-0.39	-3.95
2004-05	8.84	-0.64	-6.75
2005-06	8.74	-0.1	1.13

Source: CEA 2006.

Reductions of forced outages are deeply taken up by CEA's roving and betterment teams. The data involves 26 of 200/210 MW units, of which 19 are installed prior to March 1981. Attainment of higher PLF, reduction of outages on equipment erection faults are studied apart from other factors. Turbine problems are bearing temperatures and turbine oil system, oil pump failures, high axial shift, Vibrations in turbine, abnormal differential expansion, condenser low vacuum and tube failures. Turbine auxiliary causes also taken care. Outage leads to energy loss. The higher % of outages caused generation reduction and efficiencies. The energy losses due to Forced Outages in 2005-06 are 8.74%.

as compared to 10.79% in 1996-97. Main equipment-wise Forced Outage Loss due to Generator problems was lowest. The equipment wise break-up of total forced outage loss (as % of the All India forced outage loss) is given below:

Main Equipment	*Generation Loss as of the total Forced Outage loss*
Boiler	34.00
Turbine	17.70
Generator	6.05
Total	57.75

Out of the total number of Forced Outages of 6621, outages on account of Boiler had been the highest (2835 Nos.). The variations in before and after reforms are 16.5% after reforms 8.74%. This made possible with special efforts on boiler and turbine maintenance improvements closely analyzing failures.

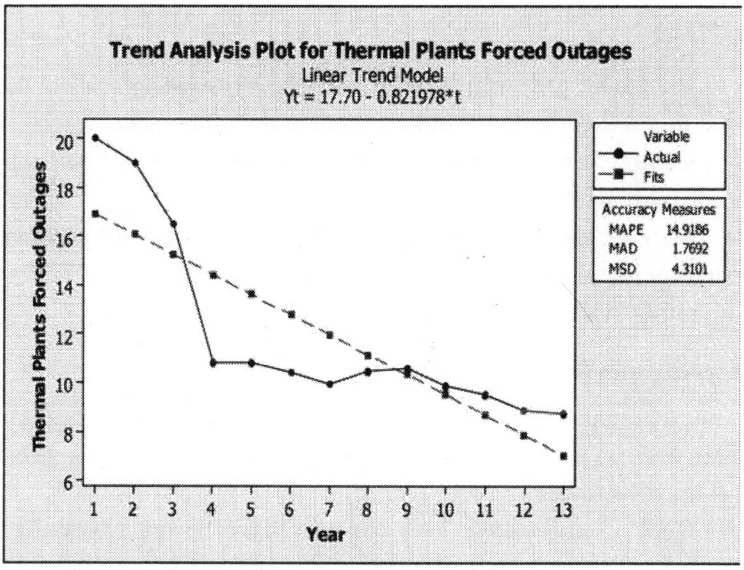

Figure 8 : *Forced Outages*

Operating Availability-(Sector wise)

Another important performance indicator. It measures the percentage of time, the plant is available for generation.

The Sector wise Planned Maintenance (%), Forced outages (%) and operating availability (%) for the last three years are as under:

Table 18
Operating Availability-(Sector wise)

Sector	*P. M. (%)*			*F.O. (%)*			*Op. Av. (%) (100-PM-FO)*		
	4 March	*5 April*	*6 May*	*4 March*	*5 April*	*6 May*	*4 March*	*5 April*	*6 May*
Central	8.35	7.03	6.85	6.86	5.88	6.37	84.79	87.1	86.78
State	9.04	9.49	11.73	10.68	10.4	10.75	80.27	80.11	77.52
Private	4.9	2.18	3.34	3.32	4.95	3.57	91.77	92.87	93.09

Source: CEA.

Private sector with 93.09% stood best operating availability due to good equipments upkeepment and maintenance.

Central sector with 86.78% availed a middle status of operating availability with moderate plan schedules.

State sector with its own efficient plan lacuna and not keeping equipments in good and suitable working conditions, have 77.52% operating availability.

Capacity wise Group Availability

The different capacity groups wise Planned Maintenance (%), Forced outages (%) and operating availability (%) for the last three years are as under:

62.5/67 and above MW capacity have better availability with regard to capacities and the operating availability percentage.

Table 19
Capacity wise Group Availability

Capacity Group (MW)	*Unavailability (%)*						*Op. Availability (%)*		
	Planned			*Forced*					
	4 March	*5 April*	*6 May*	*4 March*	*5 April*	*6 May*	*4 March*	*5 April*	*6 May*
500	6.88	5.55	7.49	4.62	4.32	5.3	88.5	90.13	87.21
250	3.09	3.58	3.15	6.29	4.88	7.62	90.62	91.54	89.23
200/210	6.03	6.95	6.76	6.05	5.81	6.8	87.92	87.24	86.44
140/150	11.77	4.39	14.29	18.35	11.12	10.28	69.88	84.49	75.43
120	12.93	13.67	5.73	34.66	24.97	30.8	52.41	61.35	63.47
110	12.74	12.02	17.02	20.35	20.86	20.33	66.92	67.12	62.65
100	18.57	20.21	23.97	14.12	19.23	9.21	67.31	60.56	66.82
70/85	4.99	6.99	5.97	21.6	21.54	9.95	72.41	71.48	84.07
62.5/67.5	11.05	5.5	11.35	9.35	8.8	10.86	79.6	85.7	77.79
60	34.7	18.11	38.13	8.91	24.43	7.56	59.39	57.46	54.31
50/57.5	42.36	46.4	49.46	11.78	7.95	8.18	45.85	45.64	42.36
20/40	12	12.28	11.15	37.7	36.37	26.83	50.3	51.44	62.02
All India	8.59	8.23	9.48	9.48	8.84	8.74	81.93	82.93	81.78

Source: CEA 2006.

Operating Availability and Unavailability

The unavailability due to forced and planned shutdown and corresponding operating availability of thermal units of 20 MW and above capacity from 1996-97 onward are shown below:

Table 20
Operating Availability and Unavailability

Year	*Unavailability (%)*			
	Planned	*Forced*	*Total*	*Operating Availability (%)*
1996-97	10.08	10.79	20.87	79.13
1997-98	10.43	10.79	21.22	78.78
1998-99	10.71	10.41	21.12	78.88
1999-2000	10.21	9.90	20.11	79.89
2000-2001	9.71	10.45	20.16	79.84
2001-2002	9.52	10.56	20.09	79.91
2002-2003	8.30	9.87	18.17	81.83
2003-2004	8.59	9.48	18.07	81.93
2004-2005	8.23	8.84	17.07	82.93
2005-2006	9.48	8.74	18.22	81.78

Source: CEA 2006.

The unavailability % has shown a decrease in 1996-97 to 2005-06 very marginally. Improvement in operational standard with respect to all equipments, raw materials, fuel etc bears the efficiency. Deficient areas of performance ant improvement needed plan. Better utilization of turbine and increasing out put of turbine were lacked. Boiler utilization and maintenance are found lagging. The operation availability in 1981-82 to 1983-84 ranges 68 to 71% in pre reforms period.

Operational Availability

Operational Availability of power plant are given year wise for after reforms period.

Table 21
Operational Availability

Year	*Operating Availability %*	*Variations*	*Percentage*
Pre reforms period			
1981-82	68		
1982-83	68	0	0
1983-84	71	3	4.41
Post reforms period			
1996-97	79.13	8.13	11.45
1997-98	78.78	-0.35	-0.44
1998-99	78.88	0.1	0.13
1999-00	79.89	1.01	1.28
2000-01	80.09	0.2	0.25
2001-02	79.91	-018	-0.22
2002-03	81.83	1.92	2.40
2003-04	81.93	0.10	0.12
2004-05	82.93	1	1.22
2005-06	81.78	-1.15	-1.38

Source: CEA 2006.

Operational Inputs to Thermal Power Stations during the Year 2005-06: Reported Loss of generation due to shortage of coal by 6 power stations was 1653.3 MU. Reported Loss of Generation due to shortage of gas by 6 Central sector, 11 State Sector, 9 Pvt. Utilities/IPPs were 7694 M. Units. However it has also been worked out that had there been no short supply of gas, the units operating on gas could have achieved 90% PLF and in this case the generation loss would have been as high as 25.53 Billion units.

Gas Turbines Performance during the Year 2005-06

Installed capacity of Gas turbine units (Utilities) under review at the end of the year are 8211.9 MW constitutes about 11% of the total thermal capacity in India. Energy generations by these gas turbines are 44078 MU. The All India Kwh/Kw/Year is 5367.5 approximately. Gas turbine stations having combined cycle operation are being given priority due to energy conservation. The installation of steam turbine for close cycle operation proves economical. The operation availability in 1996 to 2005-06 ranges 78.78 to 82.93% in post reforms period. There is a substantial difference between pre reforms 68-71% to 78.87-82.93%. This has very much improved in post reforms period.

During 2005-06 operating availability are at 81.78% decreases over 82.93 % of 2004-05%. Last five years see an increase from 79.84 to 81.78%.

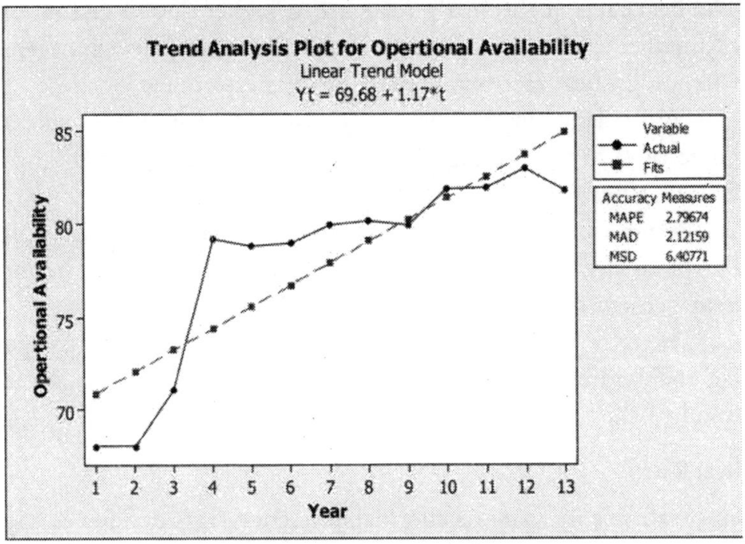

Figure 9 : *Operating Availability after Reforms*

Specific Coal Consumption

Specific coal consumption is the consumption of coal for the generation of power per Kwh. This varies state to state stations. Type of boiler, instrumentation and control system play importance. Year wise specific coal consumption in thermal plants are given below:

Energy audits of CEA reveal the following. Poor condenser vacuum that adversely affects work done per Kg of steam. Decrease in main steam and reheat steam temperatures that result in decrease in heat drop adversely effecting efficiency of stem turbine.

Part load operations resulting in reduced available heat in steam and reduced efficiency. Excessive flow of (DM) dematerialized water resulting in detoriation of turbine heat cycle rate. High combustible materials in ash due to poor grinding or incorrect combustion or incorrect pulverize classifier setting resulting in lower boiler efficiency. High flue gas temperature. All India (57 Station) average specific coal consumptions are 0.70Kg/Kwh. It is the lowest in southern region (0.65) and highest in eastern region. The heat rates at the yearly years are at higher due to factors of efficiencies. Use of source of coal, consumption pattern, technology, heat recovery system, heating system etc.

Performance in specific coal consumption in 1960's give a range of 0.76 to 0.82 Kg per Kwh. as compared to 2005-06 as 0.700 Kg per Kwh

Better efficiencies of heat rate and coal consumption are on the plan. Techniques in thermal usage and reductions, better management system are contineousely developed.

The over all specific Coal Consumption (SCC) percentage are analyzed using 13 years SCC in percentage showing a linear trend of the following type.

Heat Rate

Heat rate is a measure of thermal efficiency, (It is defined as the energy in kilo calories (kcal) required to produce a kilowatt-hour

Table 22
Specific Consumption of Coal

Year	*Specific Coal Consumption Kg/Kwh*	*Variations*	*Percentage*
Pre reforms period			
1961-62	0.762		
1962-63	0.769	0.007	0.92
1963-64	0.822	0.053	6.89
Post reforms period			
1996-97	0.73	-0.092	-11.20
1997-98	0.73	0	0
1998-99	0.72	-0.01	-1.37
1999-00	0.71	-0.01	-1.39
2000-01	0.71	0	0
2001-02	0.7	-0.01	-1.40
2002-03	0.71	0.01	1.43
2003-04	0.7	-0.01	-1.41
2004-05	0.71	0.01	0.01
2005-06	0.7	-0.01	-1.41

Source: CEA 2006.

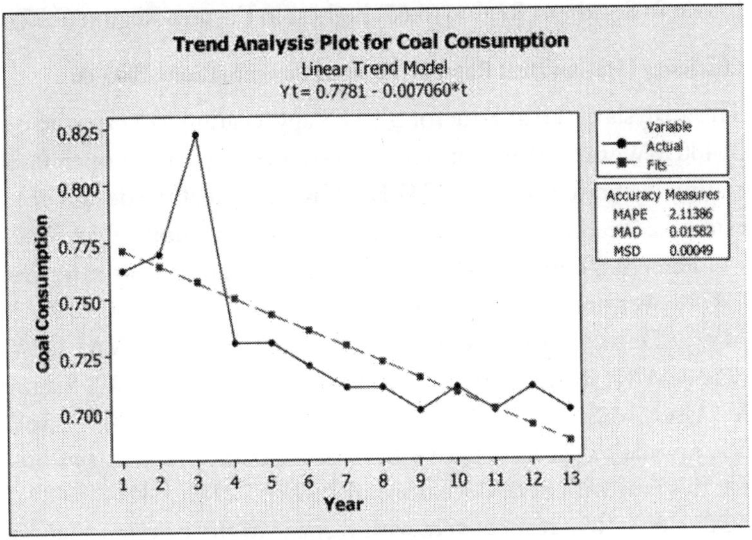

Figure 10: *Specific Coal Consumption*

(KWh) of electricity. Lesser heat rate signifies lower energy requirement to produce the same unit of electricity and hence better efficiency) a number of factors influence the heat rate of a generating plant. In the planning stage, the choice of technology, size, and fuel puts a limit on the thermal efficiency achievable. Other factors like vintage of the plant also play a role in determining the upper limit of efficiency. This heat rate is specific to each of the generating units and is known as the design heat rate. The equipment manufacturer supplies the design heat rate. In practice, it has been found that actual heat rate is higher (i.e. less efficient) than the designed heat rate. Plants using lignite have heat rates over 2500 kcal per kWh and those using LSHS have heat rates between 2700 kcal per kWh to 3400 kcal per kWh. Recent technologies based on gas or naphtha has comparatively lower heat rates ranging from 1900 kcal per kWh to 2200 kcal per kWh. In 1990, prior to the reforms, solid fuel plants having both the lowest (2300-2600 kcal/kWh) and next lowest (2600-2900 kcal/kWh) thermal efficiency are in existence. All India specific Coal consumption are 0.70 Kg/Kwh , Specific Coal Consumption lowest in Southern Region (0.65) highest in Eastern Region (0.72).

Efficiency (Station Heat Rate) of Thermal Power Stations 2005-06

The analysis of Heat Rate for 57 TPS aggregating to a capacity of 35480 MW was carried out. The analysis revealed that the operation station heat rate had been 2747 Kcal/Kwh against the design heat rate of 2398 Kcal/Kwh having percentage deviation of 14.57%.

Percentage improvement in operation station heat rate was 1.47% over preceding year. Southern Replacement CESC (2*67.5 MW), Dahanu Private Sector (2*250MW), I B VALLEY (2*210MW), Rayal Seema (2*210MW), K'Gudem (NEW), Stage-V (Unit 1&2: 2*250MW), North Chennai (3*210MW) and Vijayawada (6*210MW) have been assessed as the best stations for this year with SHR deviations of 2.69%, 3.21%, 3.41%, 3.48%, 3.92% & 4.36% respectively with respect to their design Station Heat rate (SHR).

Kutch Lignite 215MW (2x70+1x75) has improved the most during this year by improving upon its SHR by 12.9% over the preceding year, Bokaro 'B' DVC 630MW (3x210), Chandrapura DVC 780MW(3*140+3*120) and Panipat 1360MW (4*110+ 2*210+2*250) have also improved their Station Heat Rate during this year.

However, all these four stations, in spite of some improvement in their SHR over the earlier year, still continue to be amongst the most poorly operating stations with SHR deviation of 17.26%, 40.18%, 40.14% & 26.33% respectively over their design.

Poorly operating stations during this year are Barauni 320 MW (Unit Nos. 4 to 7: 2x50+2x110), Paricha 220MW (2x110), Amar Kantak EXT, 240MW (Unit Nos. 1 & 2:2x120), Faridabad Extn. 180 MW (3x60), Santaldih 480 MW (4x120), Neyveli ST-I 600 MW (6x50+3x100) and I.P. Station 247.50 MW (Unit Nos. 2 to 5:3*62.5+1*60) with SHR deviations of 109.81%, 69.96%, 66.71%, 49.82%, 48.71%, 45.73% & 41.4% respectively over their design Station Heat Rate.

Partial Loss Percentage (Due to Equipment)

Partial unavailability are caused by non repetitive stages of turbine blades, stabilization problems of new units, non completion of control/instrumentation requirements all accounts roughly 6%.

Repetitive constraints in mill, fans, variation in quality of coal etc. accounts roughly 10%. Readiness of main equipments and auxiliary equipments causes higher loses during these period. Pre reforms partial loss % are 1981-85 are 14-16%.

Causes of Unavailability

Various causes of unavailability are given below:

Unavailability due to internal constraints arising out of deficiency in generating units/auxiliaries was 7.08%. RUSSIA/ RUSSIA make units had lowest Partial Unavailability of 0.33% and highest Partial Unavailability was in BHEL/BHEL group i.e.5.31%.

Table 23
Causes of Unavailability

Constraint Area	*MU Loss*	*Main Cause*
Boiler	2708.3	Pressure Parts Leakage/drum pressure Restriction
Boiler Aux	8293.8	Milling System
Turbine	3914.7	Startup/shutdown
Turbine Aux	1477.6	Regenerative system
Generator	94.7	Hydrogen pressure/Purity
Misc. Elec. Cause	3341.3	Grid System
Misc Mech./General Cause	21101.8	Poor quality/wet coal
Low sys. Demand	7497.0	Low sys. Demand

Source: CEA 2006.

Partial Loss Percentage (Due to Equipment)

The All India loss of generation due to Partial loss because of Main Equipment is given.

Table 24
Partial Loss Percentage (Due to Equipment)

Year	*Partial Loss % Due to Equipment*	*Variations*	*Percentage*
Pre reforms period			
1981-82	16		
1982-83	14	-2	-12.5
1983-84	14	0	0
Post reforms period			
1996-97	13.25	-0.75	-5.36
1997-98	12.4	-0.85	-6.42
1998-99	12.04	-0.36	-2.90
1999-00	11.04	-1	-8.30
2000-01	9.88	-1.16	-10.51
2001-02	9.99	0.11	1.11
2002-03	9.55	-0.44	-4.40
2003-04	9.09	-0.46	-4.82
2004-05	8.35	-0.74	-8.14
2005-06	8.38	0.03	0.35

Source: CEA 2006

The All India loss of generation due to main Equipment was 1.16% and loss of generation due to Partial Unavailability because of Auxiliaries was 1.69% of the overall loss of generation. Post reforms partial loss % are 1996-97 to 2005-06 are 13.25 - 8.38%. as compared to Pre reforms partial loss % of 14-16%. This is a good reduction between pre and post reforms period. Partial Loss (PL) percentage are analysed using 13 years PL in percentage showing a linear trend of the following type. $Vt = a + bt$ Where Vt is PL in year t, t varying from 1 to 13 years and a and b are constants is fitted to the yearly average PL . The Partial Load % has decreased over the years. The decreasing trend value is Yt = 15.678-0.613681*t .

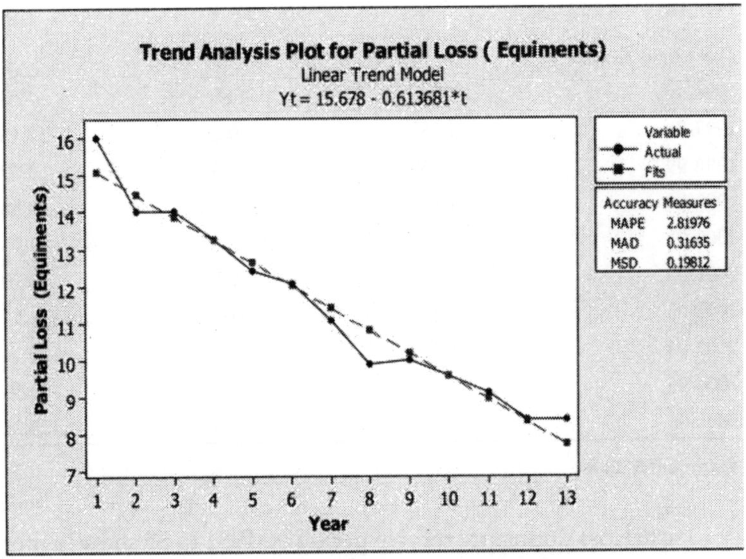

*The decreasing trend value is Yt = 15.678 - 0.613681*t*

Figure 11: *Partial Loss Percentage (Due to equipment): After Reforms*

Partial Loss Percentage (Due to Low System Demand/Resrve shutdown)

Less generation by operating units are due to system load variation (Low system Demand). Low load or non utilization of available units are due to their complete shutdown and for want of load

(Reserve shutdown). Year wise partial loss is an efficiency factor given below for LSD/RSD causes. Low load or non utilization of available units are due to their complete shutdown and for want of load (Reserve shutdown). Year wise partial loss is an efficiency factor given below for LSD/RSD causes.

Table 25
Partial Loss Percentage
(Due to Low System Demand/Reserve Shutdown)

Year	*Partial Loss % – LSD/RSD %*	*Variations*	*Percentage*
1981-82	5		
1982-83	5	0	0
1984-85	6	1	20
Post reforms period			
1996-97	1.35	-4.65	-77.5
1997-98	1.55	0.2	14.48
1998-99	2.33	0.78	50.32
1999-00	1.89	-0.44	-18.88
2000-01	1.43	-0.46	-24.33
2001-02	1.53	0.1	6.99
2002-03	1.57	0.04	2.1
2003-04	1.98	0.41	26.11
2004-05	0.92	-1.06	-53.53
2005-06	2.1	1.18	128.26

Source: CEA 2006

Partial loss during pre reform period of 1981 to 85 show higher losses ranging 5 to 6% as compared to 0.92 to 1.35 % in post reform period.

Partial Unavailability and Non-utilization of Energy Due to System Load Variation during 2005-06

- Post reforms period show a marked reduction.
- The gap between Operating Availability and P.L.F. was 8.38%

- Partial Unavailability due to low System Demand was 1.30%.
- Partial Unavailability was highest in Eastern Region (11.3%) & lowest in Southern Region (6.4%). After reforms period partial loss due to LSD/RSD ranges 2.33 to 0.92 in the years 1996 to 2006. As compared to pre reforms it is much improved.

Partial Loss-Low System Demand (PL & LSD) percentage are analysed using 13 years PL & LSD in percentage showing a linear trend of the following type.

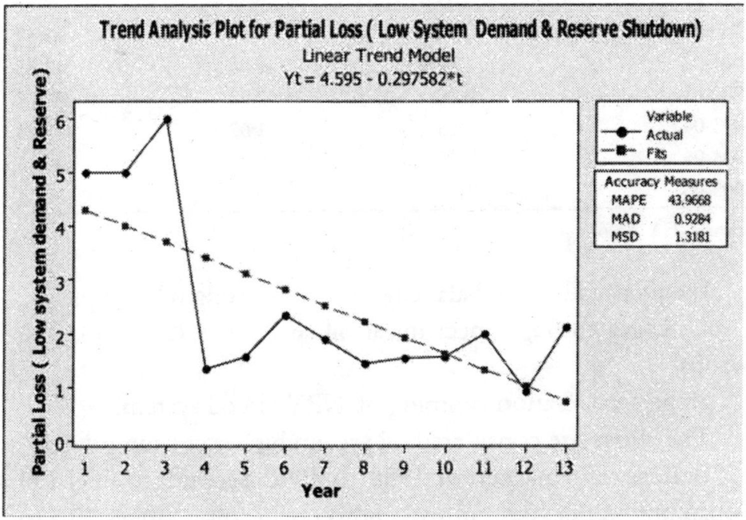

Figure 12 : *Partial Loss Percentage (Due to Low System Demand/ Reserve shutdown)*

Secondary Oil Consumption

Secondary Fuel Oil Consumption per Kwh of gross power generation are expressed in ml/Kwh. This factor vary plant to plant and one of the efficiency factor of a power plant. One per cent reduction on this will save huge cost. Year wise before reforms period are given below.

Table 26
Secondary Oil Consumption

Year	*Secondary oil Consumption Ml/Kwh*	*Variations*	*Percentage*
1988-89	10.60		
1989-90	10.30	-0.3	-2.83
Post reform period			
1996-97	4.36	-5.94	-57.67
1997-98	3.45	-0.91	-20.87
1998-99	3.5	0.05	1.45
1999-00	3.59	0.09	2.57
2000-01	2.75	-0.84	-23.39
2001-02	2.7	-0.05	-1.82
2002-03	0.68	-2.02	-74.81
2003-04	2.3	1.62	238.23
2004-05	1.37	-0.93	-40.43
2005-06	1.77	0.44	32.11

Source: CEA 2006.

Frequent failure of balancing disc of starting oil pumps.

Leakages through mechanical oil seals of A.C and D.C oil pumps.

Sluggish operation jamming of NRV,s in oil system.

The above are considered oil system higher consumptions.

Before reforms period 1988 to 1990 Secondary Fuel Oil Consumption per Kwh of gross power generation ranges 10.3 to 10.60 ml/Kwh. This is on the higher side, and need reduction of consumption.

Year wise after reforms period are given below. All India (57 Stations) Average Specific Secondary Fuel Oil Consumption was 1.77 ml/Kwh. in 2005-06. The consumption pattern has come down during post reforms period considerably due to new methods and control affected. in flow of oil, primary leakages, developed control in quantity usages and replacement of more efficient oils that are brought over a period . In the year 2003-04 the variation is high consumption due to un found reasons and control failures.

The over all Secondary Oil Consumption (SOC) percentage are analysed using 12 years SOC in percentage showing a linear trend of the following type.

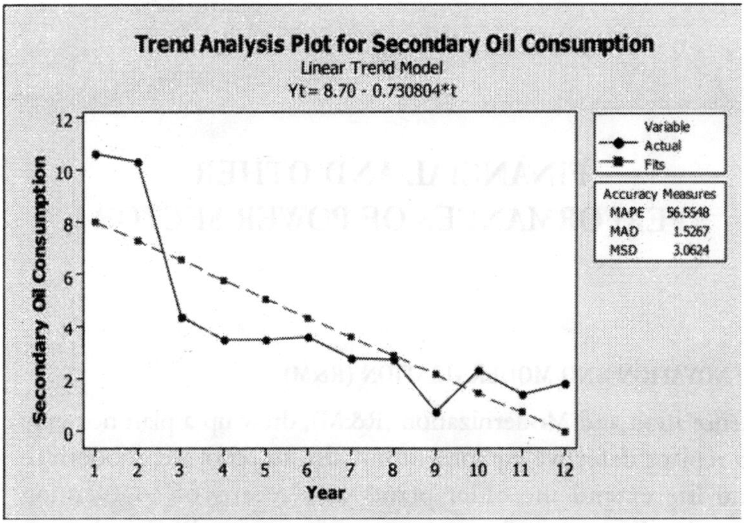

Figure 13: *Secondary Oil Consumption*

CONCULSION

The operating performances of various plants pre and post reforms has shown marked operational parameters and improved their operating efficiencies leading to healthy plant operations. In 2006-07, 66 units had PLF more than 90% and one unit reported more than 100%, and over all PLF stood at 77%. Operating availability stood at 83.7% in 2006-07. Specific coal consumption for 2006-07 was at 0.71%. Generation grow in 2006-07 by 6% to last year. No unit made loss of shortage of coal in 2006-07, showing better coal linkage. Many units have crossed 25 years life span require R&M activity for better performances. Otherwise the year of 2006-07 show impressive improved performance in its operations.

4

FINANCIAL AND OTHER PERFORMANCES OF POWER SECTOR

RENOVATION AND MODERNIZATION (R&M)

Renovation and Modernization (R&M), drew up a plan not only to replace defective equipment but also to renovate, modernize and life extend the older plants as a means of augmenting capacity. Economically improve performance, PLF and more generation.

In the seventh, eighth, ninth and tenth R & M plans, as per above Table 1, an additional power generation gained shown are 53550 Million Units, due to R&M besides PLF % gains, and life got extended for the plants. Seventh plan show an increase of 100% variation with -31% in 9th plan due to slump in the programme but the 10th plan see an increase in variation of 290% due to the vigorous implementation of the programme. R&M has gained additional generation of power from old plants at a less investments. Otherwise new power plants installations and to gain this power generation require much of investments, time and more efforts.

PER-CAPITA ELECTRICITY CONSUMPTION IN INDIA

Energy consumption as compared to developing country lags very much. India maintains a low per-capita power consumption of

Table 1
R&M Achievements: Before and after Reforms– 7th, 8th, 9th, 10th Plans

Plan	*No. of Units*	*Variation*	*Percentage*
Additional generation achieved 87-92 7th plan	10000	10000	100
Additional generation achieved 92-97 8th plan	10000	0	0
Additional generation achieved 97-02 9th plan	6850	-3150	-31.5
Additional generation achieved 02-07 10th plan	26700	19850	290
Total	53550		

Source: CEA & MOP 2001, 2007.

366.1 Kwh. in 2001. In the Growth of Indian per-capita capacity consumption thermal generation are substantial where as other sources are very meager share. Year wise per-capita consumption are given below.

Table 2
Year wise Electricity Per-capita in India

Year	*Per-capita Kwh*	*Variation*	*Percentage*
Pre reforms period			
1950	15.6		
1956	26.4	10.8	69.23
1961	37.9	11.5	43.56
1966	61.4	23.5	62.00
1969	77.9	16.5	26.87
1974	97.5	19.6	25.16
1979	130.9	33.4	34.27
1980	130.5	-0.4	-0.31
1985	168.5	38	29.11
1990	238	69.5	41.25
Post reforms period			
1992	270	32	13.44
1997	334.3	73.3	27.14
2001	366.1	31.8	9.51
2002	559.2	193.1	52.83
2005	606.2	47	8.40
2006	616	9.8	1.62

Source: Power Line Dec. 2005 & CEA 2006.

The power per-capita consumption was very low with 238 KWh, during pre reforms period that attributed to low generation and lesser power infrastructures. The power per-capita consumption are higher during post reforms period that attributed to higher generation and more power infrastructures development. Power requirement for different consumer category have increased. Agriculture, commercial, industry, domestic consumptions have increased. Personal power equipments and house hold power usage have multiplied are the reasons. Industrial consumptions have multi fold increased. Transportation by rail, mechanization of work started using more power all these are met from thermal power generations which accounted more than 81 % of total generations. Year wise variations in 1990 is high 69.5KWh/year in the pre reforms period are due to higher power generations in thermal sector. In the year 2002 see a high variation of 193.1 kWh/year are also due to higher power generations in the thermal sector.

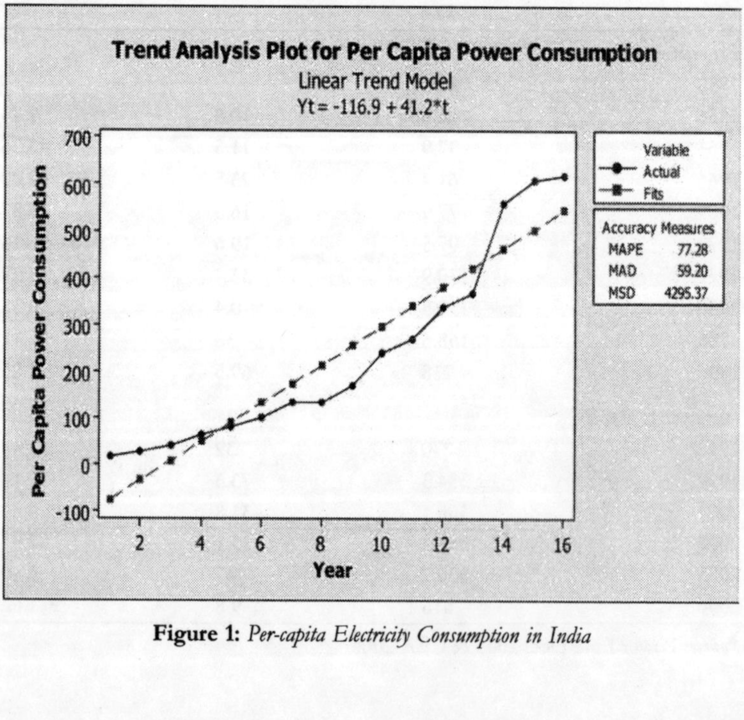

Figure 1: *Per-capita Electricity Consumption in India*

SHIFT IN CONSUMPTION PATTERN

Shift in before and after reforms period Consumption % are given above.

Decade wise domestic, commercial, industry, agriculture consumers percentage of consumption are given from 1960 to 1991 as per Table 3. The shift in industry and agriculture are largely due to increase in service connections, more availability of power and cheap tariff for agricultural sector.

Table 3
Shift in Consumption Percentages

Year	*Domestic*	*Commercial*	*Indu stry*	*Agriculture*	*Others*
1960-61	10.7	6.1	69.4	6	7.8
1970-71	8.8	5.9	67.6	10.2	7.5
1980-81	11.2	5.7	58.4	17.6	7.1
1990-91	16.5	6.1	44.9	26	6.7

Source: MOP & CEA 2001.

INSTALLED CAPACITY

The installed capacity in Indian power sector generations has grown to 1,24, 287 MW as on 31.03.2006 since 1950 from 1713 MW are given.

Above do not include captive capacities of 25000 MW. India has a potential growth in all four areas- Thermal, hydro, Nuclear and renewable energies. New power installations are the only way to meet demand and shortages.

Installed capacity as as per Table 4, has grown from 107 MW in 1915 to 124287 MW in 2006. A growth of 124180 MW in 91 years. A centaury of growth, now required to be increased in next eleventh plan to meet power shortages. The growth acceleration since 1915 has seen different variations. A 722% growth between 1915 to 1945 in 27 years averaging 27% per year as per base year 1915 capacity during this period. 1956 to 1966 period has a growth of 97.87% averaging 9.8% per year. 1974-79 see a growth of 60.105 in 5 years averaging

Table 4
Installed Capacity

As On	*Utilities*				*(In MW)*
	Thermal	*Hydro*	*Nuclear*	*Wind*	*Total*
1915	4	103	—	—	107
1918 (End Of First World War)	27	103	—	—	130
1945 (End Of 2nd World War)	627	442	—	—	1069
31-12-1947	854	508	—	—	1362
31.12.1950	1153	559	—	—	1713
31.03.1956(End Of First Plan)	1825	1061	—	—	2886
31.03.1956(End Of second Plan)	2736	1917	—	—	4653
31.03.1966(End Of third Plan)	4903	4124	—	—	9207
31.03.1969(End Of 3annual Plan)	7050	5907	—	—	12957
31.03.1971	7900	6400	400	—	14700
31.03.1974(End of the 4th plan)	9058	6965	640	—	16664
31.03.1979(end of the 5th plan)	15207	10833	640	—	26680
31.03.1980 (end of the annual plan)	16424	11384	640	—	28448
31.03.1981	17600	11800	640	—	30000
31.03.1985 (end of the 6th plan)	27030	14460	1095	—	42585
31.03.1986	30000	15500	1095	—	46460
31.03.1990 (end of 2 annual plans)	43764	18308	1565	—	63636
31.03.1991	45768	18753	1565	—	66086
31.03.1992	4805	19194	1785	41	69065
31.03.1993 (end of 2 annual plans)	50695	19576	2005	54	72330
31.03.1994	54254	20379	2005	115	76753
31.03.1995	57762	20833	2225	351	81171
31.03.1996	59334	20976	2225	733	83288
31.03.1997 (end of the 8th plan)	61010	21658	2225	902	85795
31.03.1998	645005	21904	2225	968	89102
31.03.1999	67534	22497	2225	1024	93280
31.03.2000	70201	23816	2680-	1155	97837
31.03.2001	72344	25153	2680	1269	101626
31.03.2002 (end of the 9th plan)	74429	26261	2720	1628	104917
31.03.2003	76607	26910	2720	1736	107973
31.03.2004	77958	27560	2720	1950	110188
01.01.2005	79451	30080	2720	2488	114739
31.03.2006	82410	32325	3360	6191	124287

Source: CEA & MOP -2006.

Table 5
Installed Capacity-with Year Wise Variations

As On	*(In MW) Total*	*Variations*	*Percentage*
1915	107		
1918 (End Of First World War)	130	23	21.49
1945 (End Of 2nd World War)	1069	939	722.30
31-12-1947	1362	293	27.41
31.12.1950	1713	351	25.77
31.03.1956 (End Of First Plan)	2886	1173	68.47
31.03.1956 (End Of second Plan)	4653	1767	61.23
31.03.1966 (End Of third Plan)	9207	4554	97.87
31.03.1969 (End Of 3annual Plan)	12957	3750	40.73
31.03.1971	14700	1743	13.45
31.03.1974 (End of the 4th plan)	16664	1964	13.36
31.03.1979 (end of the 5th plan)	26680	10016	60.10
31.03.1980 (end of the annual plan)	28448	1768	6.63
31.03.1981	30000	1552	5.46
31.03.1985 (end of the 6th plan)	42585	12585	41.95
31.03.1986	46460	3875	9.10
31.03.1990 (end of 2 annual plans)	63636	17176	36.97
31.03.1991	66086	2450	3.85
31.03.1992	69065	2980	4.51
31.03.1993 (end of 2 annual plans)	72330	3265	4.73
31.03.1994	76753	4423	6.12
31.03.1995	81171	4417	5.75
31.03.1996	83288	2117	2.61
31.03.1997 (end of the 8th plan)	85795	2507	3.01
31.03.1998	89102	3307	3.85
31.03.1999	93280	4178	4.69
31.03.2000	97837	4593	4.92
31.03.2001	101626	3789	3.87
31.03.2002 (end of the 9th plan)	104917	3291	3.24
31.03.2003	107973	3056	2.91
31.03.2004	110188	2215	2.05
01.01.2005	114739	4551	4.13
31.03.2006	124287	9548	8.32

Source: CEA & MOP-2006.

12% growth. 1981-85 see a growth of 41.95% averaging 8.395 per year to that period. 1986-90 see a growth of 36.97% averaging 7.395 per year. Last year 2006 see a growth with 8.32 % against all sect oral difficulties. In the capacities added Thermal power which accounts for more than 66.34% in MW. However MOP see to that a stern progress is achieved to make up power shortages. The growth since 1915 to 2006 is progressive but not very aggressive. Growths of power plant installations since 1915 are marginal. In 75 years it is 63500 MW. The growths of power installation since 1991 are very high. It is 58201 MW in just 15 years. The power sector's major demands of power are met by huge thermal capacity addition only.

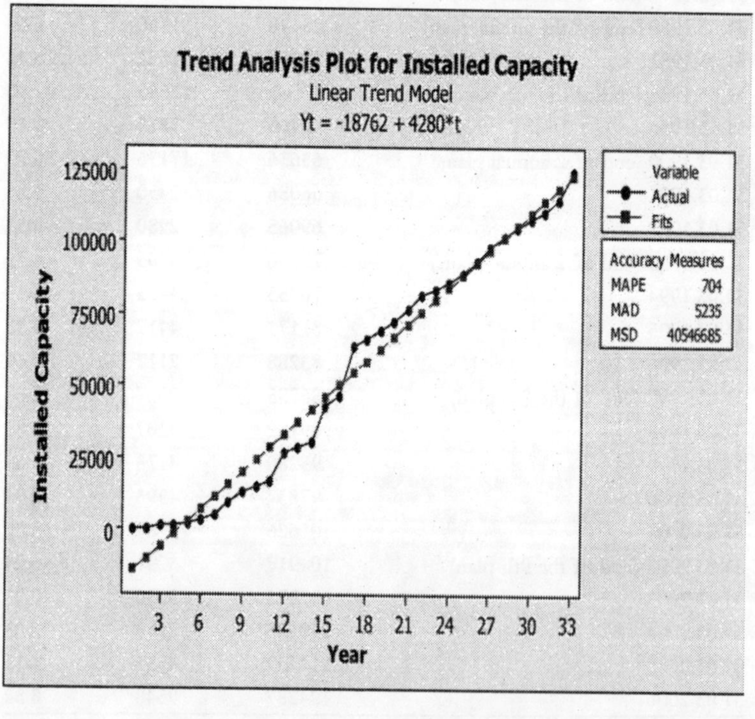

Figure 2 : *Installed Capacity Additions in MW 1915-2006*

GROWTH OF INSTALLED CAPACITY OVER PLAN PERIOD WISE

Table 6

Plan	*End of plan*	*Installed MW*	*Variations*	*Percentage*
I	31.03.56	2695		
II	31.03.61	4653	1958	72.65
III	31.03.66	9027	4374	94.00
IV	31.03.74	16664	7637	84.60
V	31.03.79	26680	10016	60.10
VI	31.03.85	42585	15905	59.61
VII	31.03.90	63633	21048	49.42
VIII	31.03.97	84893	21260	33.41
IX	31.03.02	104917	20024	23.59
X	31.03.07 Upto 01.01.07	127925	23008	51.52

Source: MOP 2007.

Plan wise capacity additions of power are given. Various plan periods from I to Xth plans up to 01.01.2007 have brought 127 925 MW in the Indian power sector to cater to the power sector and the nation. First four plan periods have slow growth. Fifth plan to tenth plan a growth momentum picked up to meet demands of power and due to government's accelerated installation plan programmes. XI th plan set a target of 78,577 MW. In this 16,335 MW for 2007-08 targetted, but 9263 MW actual achieved (56.71%), leaving 43.29%. This is attributed to shortage of equipments, delys, land acquisition and rehabilitation problems.

GROWTH OF INSTALLED CAPACITY ADDITION OVER PLAN PERIOD WISE

The Multiplicative method explains the trend value for growth of installed capacity over the plan periods in Indian Power sectors. The trend shows that over the (five year) plan periods, growth of installed capacity have steeply increased as trend which it reveals, growth of installed capacity increases to meet power consumption of growing population in India.

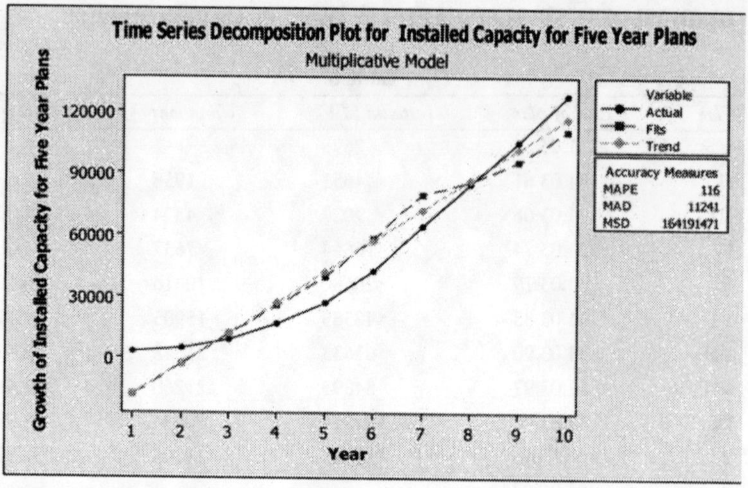

Figure 3

GROWTH OF INSTALLED CAPACITY ADDITION OVER PLAN PERIOD WISE-1

Above figure shows that government of India is expending for growth of installed capacity additions in five year plans.

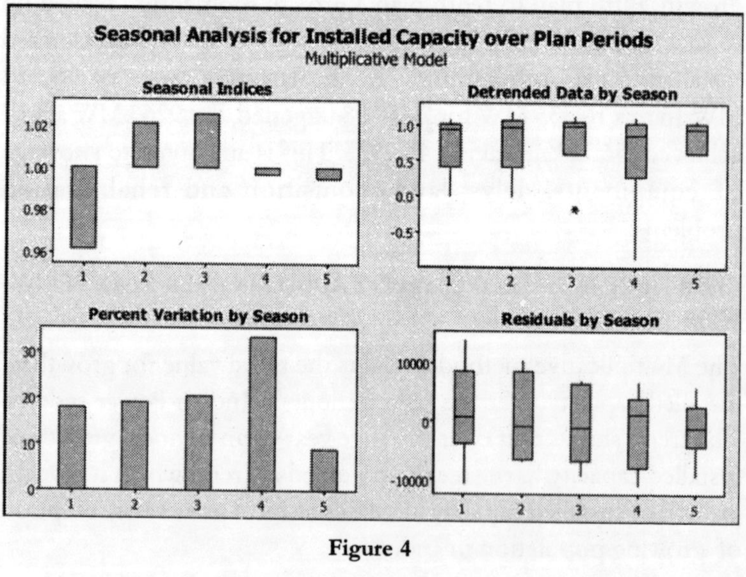

Figure 4

GROWTH OF INSTALLED CAPACITY ADDITION OVER PLAN PERIOD WISE-2

The figure explains that during first three five year plans, the government made less steps to increase installed capacity for promoting power generation capacity. After fourth plan, the government recognized that the power is very important for human life. It also states, during the fourth plan, country faced urbanization and require more power for consumption. After the fourth to nine plan, a substantial increase in installation are proposed and carried out.

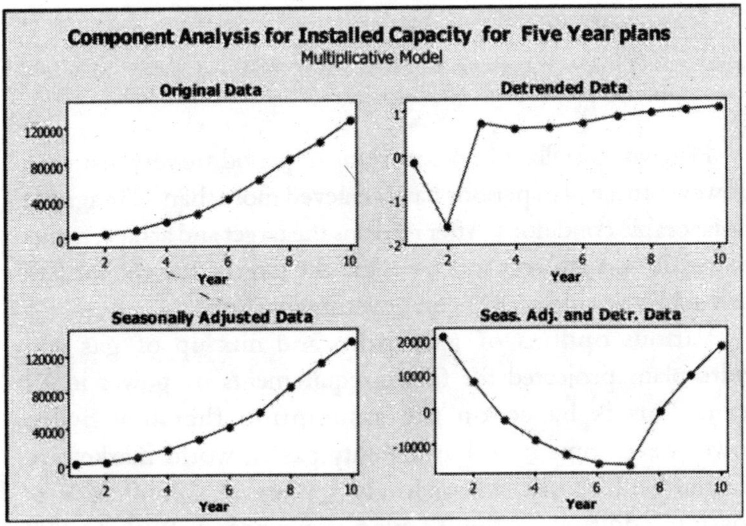

Figure 5

PLAN WISE TARGETS AND ACHIEVEMENTS

Installed capacity of power plants are 127287 MW as on 31.03.2006. At present a capacity of 34024 MW is likely to be achieved during Xth plan ending 31.03.2007 as per Ministry of power and CEA' mid term appraisals.

Almost 5000 MW hydro projects lacked preparedness and serious flaw in planning if such projects are added in the plan. In 2006-07 as per MOP about 20,000 MW is likely to be added.

Measures are made to salvage the situation for national capacity addition programme.

Table 7
Plan Wise Targets and Achievements

Plan	*Period*	*Target*	*Achieved*	%
V	1974-79	12499	10202	81.6
VI	1980-85	19666	14226	72.3
VII	1985-90	22245	21401	96
VIII	1992-97	30538	16422	53.7
IX	1997-2002	40245	19015	47
X	2002-07	41110	21180*	51.52

Source: MPO 2006. *Achievement projected by MOP and include 2520 MW nuclear project under construction.

Plan wise installations before reforms period are very marginal. However these plan periods have achieved more than 72% against un-favorable conditions. After reforms the target and achievements on installation are very much behind the targets. Except for Xth plan achievements of 82% on government thrust.

Various options of gas, hydro and mix up of gas and hydro-plans projected for future requirements of power in XI plan. This is based on the assumption that low hydro power development and availability gas in world market are limited and costly. Accordingly a plan of 68500 MW is possible addition in the XI plan is proposed. With 65,000 MW already placed orders for capacity creation under XI plan, MOP is confident of all orders completion by end of XI plan period as per Power Minister's statement reported to press on 16.02.2008.

GDP GROWTH PROPOSED INSTALLATION CAPACITY OF POWER IN MW CAPACITY FOR 10TH PLANS TO 14TH PLANS

Tentative targets for capacity addition during 10th-14th plans based on 8 % GDP taking into various potentials of thermal, hydro, pump storage, nuclear power fronts.

Table 8
Proposed Installation Capacity-10th Plans to 14 th Plans

End of	Year	*Thermal* Coal / lignite	Gas/ oil	Nuclear	*Medium & large hydro in India* Pump storage	Hydro power	*Import from* Bhutan & Nepal	*Other renewable energy sources for the grid* Small Hydro	Wind	Others	Total
9th plan (Actual)	2001-02	62131	12298	2720	1554	23287	—	1420	1508	—	104918
10th plan	2006-07	82184	17662	4020	3704	35530	1020	1820	2508	400	148848
11th plan	2011-12	106184	23662	10020	5704	55530	2020	3120	4508	1300	212048
12th plan	2016-17	133684	30662	18020	8204	79530	3120	4620	6908	2300	287048
13th plan	2021-22	167684	38662	30020	11204	105530	4820	6220	9508	3400	377048
14th plan	2026-27	217184	47662	50020	15204	134530	6720	8020	12908	4800	497048

Source: CEA.

Table 8 explain proposals based on GDP growth at 8%. IX plan of 2007-12 should have a capacity installed 212048 MW as present target to avoid future shortages. Government has seen a slump in 9th and 10 th plans and developed a three stage monitoring system planned are, overall master control at top level, package wise scheduling at level II and milestone oriented detailed scheduling at working level III is periodic review by MOP through CEA.

LAUNCH OF ULTRA-MEGA POWER PROJECTS (UMPP'S)

Recognizing that economies scale leading to cheaper power can be secured through development of large size power projects, ministry of power, CEA and power finance Corporation have worked out Ten ultra mega power project under tariff based competitive bidding route. These projects will be awarded to developers on Build own and Operate basis. Each project are with a capacity of 4000 MW, would have cope for further extension. As the projects are bigger they will be able to meet power needs of a number of states. Five wholly owned subsidiaries have been established by Power Finance Corporation Ltd are:

1. Sasan Power Limited, apit head project at Sasan (M.P).
2. Akaltara Power Limited, a pit head projects at Akaltara (Chattisgarh).
3. Costal Gujarat Power Limited, imported coal project, Mundra, Gujarat.
4. Coastal Karnataka Power Limited, imported coal project in Karnataka.
5. Coastal Maharastra Mega Power Limited, an imported coal based project in Maharastra.
6. Five more projects, a pit head at Orissa and a coastal site in Andhra Pradesh and Two in Tamil Nadu are being explored.

The programme will take care of required capacity addition installations and are coal based plants. Sustaining power growth is the primary aim, and also to meet targeted power installations in the country to meet demand and supply of power. Award of three UMPP's in Mundra, Sasan and Krishnapatnam of 4000 MW to private sector in 2007-08.

RURAL ELECTRIFICATION

Rural Electrification is a vital programme for socio-economic development of rural areas. The objectives are to trigger economic development and generate employment by providing electricity as an input for productive uses in agriculture and rural industries, and improve the quality of life of the rural people by supplying electricity for lighting of homes, shops, community centers and public places in all villages.

Status: Out of 5,87,258 total inhabited villages in the country as per 1991 census, 5,08,863 villages i.e 87% villages have been declared elcctrified by March 2002. Balances of 77,897 villages still remain to be electrified. The state-wise of village of electrification in the country as on 31.03.2002 is given.

Nine States have achieved 100% village electrification.

These are: Andhra Pradesh, Goa, Haryana, Kerala, Punjab, Tamil Nadu, Maharashtra, Gujarat and Sikkim.

Seven States are about completing village electrification. These are: Nagaland, Mizoram, Karnataka, Himachal Pradesh, Jammu & Kashmir, Tripura and Manipur.

STATE WISE STATUS OF RURAL ELECTRIFICATION NOT FULLY ELECTRIFIED AS ON 2002

The position electrified status of the twelve States, according to CEA are given as under:

These States have been requested to indicate the authenticated number of un-electrified villages as on 31.03.2002.

Table 9

Rural Electrification Not Fully Electrified as on 2002

State	*Total no. of Inhabited villages*	*Villages Electrified*	*Balance*	*% Electrified*
Uttranchal	15681	12488	3193	80
Uttar Pradesh	97122	77152	19970	79
West Bengal	37910	29636	8274	78
Assam	24685	19019	5666	77
Orissa	46989	352332	11757	75
Bihar & Jharkhand	67513	47950	19563	71
Arunachal Pradesh	3649	2206	1443	60
Meghalaya	5484	2551	2933	46
Rajasthan	37889	36488	1401	95
Madhya Pradesh	51806	50286	1520	97
Chattisgarh	19720	18076	1644	92

Source: MOP 2002.

Table 10

State	*Total no. of Inhabited villages*	*Villages Electrified*	*Balance*	*% Electrified*
Andra Pradesh	26586	26565	99.9	0
Arunachal Pradesh	3649	2206	1443	60
Assam	24685	19019	5666	77
Bihar & Jharkhand	67513	47950	19563	71
	-	-	-	-
Chattisgarh	19720	18076	1644	92
Madhya Pradesh	51806	50286	1520	97
Meghalaya	5484	2551	2933	46
Orissa	46989	352332	11757	75
Rajesthan	37889	36488	1401	95
Uttar Pradesh	97122	77152	19970	79
Uttranchal	15681	12488	3193	80
West Bengal	37910	29636	8274	78
Goa	360	360	100	0
Gujarat	18028	17940	99.5	0
Haryana	6759	6759	100	0
Himachal Pradesh	16997	16886	99.3	111
Jammu& Kashmir	6477	6315	97.5	162
Karnataka	27066	26759	98.9	26
Kerala	1384	1384	100	0
Maharashtra	40412	40349	99.8	0

Contd..

State	Total no. of Inhabited villages	Villages Electrified	Balance	% Electrified
Manipur	2182	2001	91.7	181
Moghalaya	5484	2551	46.5	2933
Mizoram	698	691	99	7
Punjab	12428	12428	100	0
Sikkim	447	405	90.6	0
Tamil Nadu	15822	15822	100	0
Tripura	855	813	95.1	42
Total	586165	507425	86.6	77897
UTs	1093	1090	99.7	0
G. Total	587258	507773	86.6	77897

Source: CEA Feb., 2002.

STATUS OF VILLAGE ELECTRIFICATION IN INDIA AS ON MARCH 2002

The Central Electricity Authority (CEA) has identified out of 587258 villages, about 77897 have no supply infrastructure. About 51,284 of these have been identified for electrification by 2006-07. For electrifying these villages, 192 projects have been identified in 29 districts of 22 states. The total capital sanctioned for these projects is Rs. 62,850 million.

To complete electrification of 61,671 villages through grid supply, the estimated fund requirement is of the order of Rs. 4,580crore i.e. an annual investment of Rs. 916 crore for completion of the balance works by the end of 10th Plan i.e.2007. On an average, electrification of about 12,000 villages would require to be completed every year by the States to complete the task within the desired time frame i.e. by 2007.

RURAL ELECTRIFICATION PROGRESS AND TARGETS

Rural electrification has been given thrust to electrify villages before 2007.

By May 2006, 12,239 villages had been electrified. Uttar Pardesh and Bihar have benefited significantly from the RGGVY so far.

Table 11
Rural Electrification Progress and Targets

Electrified as on March 2006		*To be electrified by March 2007*	
State	*No. of Village*	*State*	*No. of Village*
Bihar	1,600	Bihar	10,097
Uttar Pradesh	7,503	Uttar Pradesh	22,845
Karnataka	397	Jharkhand	1718
Others*	679	Orissa	250
Total	10,169	Others*	5,250
Total reported up to May 2006	12,12239	Total	40,160

* West Bengal, Rajasthān, Uttaranchal.

Source: REC

The programme is fast paced given that another 40,00 villages are to be electrified by 2006-08. It also covers partly electrified villages.

Rural Electrification is a reform programme to develop rural infrastructure and to make villages development. Table 10. Explain in 2006, that 10169 villages are electrified. 2007 targets for 40160 villages, so that all villages in India will be electrified. Required powers are met from thermal capacity additions and from newly planned renewable or locally generated power.

FINANCIAL PERFORMANCES OF THERMAL POWER UTILITIES

Financial health of power utilities are very important factor. The utilities growth, development, expansions, R&D, progress in all direction, modernization all depends on a healthy and sound financial performances of utilities. Various thermal utilities financial performance is gauged on their profits, Rate of Return (ROR), commercial losses achieved in the process. Leading power utilities financial highlights are given below:

NTPC: (National Thermal Power Corporations Limited)

The largest thermal generator in the country incorporated in 1975, has achieved the highest ever generation of 170.8 Billion units in 2005-06, a 7.4% increase over last year and contributed 285 of nation's power generation in the year. It recorded 87.54 % PLF and achieved 100% realization. NTPC has signed 10 loans aggregating 37.6 billion rupees taking a total of Rs.195.3 billions. It also established a $1 billion medium term programme and offered the first tranche of $300 million. NTPC's trading arm traded 1.6 billion units of power in the year transacting trading with 18 utilities. Its consultancy wing secured 73 orders valued at Rs. 19 billions.

NHPC: (National Hydroelectric Power Corporation Limited)

In corporated in 1975, it registered a net profit of Rs. 7.43 billion during 2005-06 as against 6.5 billion on last year. Net sales of 16.14 billion compared to last year 14.49 billion. It generated 12.56 billion units with an increase of 11.3% over previous year.

NLC: (Neyveli Lignite Corporation Limited)

NLC a Public company incorporated in 1955, had a profit in 1986-87 of Rs. 67.8 crores (0.678 Billion), presently the net sale are 25.02 billion in 2005-06. Net sales declined to 16%, over 2004-05 and net profit also come down to Rs. 7.02 against 12.15 in the last year. As the company has adopted CERC's depreciation policy, since the same being one of the components of fixed charge in tariff, the over all tariff charged by the company will be lower and led to lower profits. Lignite mining accounts 73% of revenue and power for the balance. Profits have fallen in generations to 418 millions from Rs. 6.8 billion, in lignite it fell to 4.50 from Rs. 7.33 billion from last year.

RELIANCE ENERGY

A private company in thermal, reported a net profit of Rs. 6.5 billion, a 25% increase over last year. It had atotal income of Rs. 46.08 billions against last years 45.93 billion. Power ale revenue are 31.79 from 28.96 billion on last year. Aggregate energy sales were 8064 million units with an increase of 1.30%. It has a consumer of 2.49 millions. It purchased 3921 MU energy from outside with a 17% increase over previous year. Its Dhanu plant recorded 98.7 PLF down from 101.4% from previous year. It wind form project recorded 34% PLF.

TATA POWER

TATA powers is a private company, in hydro and power distribution company, its net profit rose 10.7% higher than last year. Income rise of 16% to 45.62 billions, and detached its last year Tata petrodyne as fully owned subsidiary. Company generated 13.74 billion units power in 2005-06, up fro 13.6 billion last years. Revenue from power supply at Rs. 42.9 billion accunted for 94% of net revenues.

***PTC*: (Power Trading Corporation India Limited)**

Power Tradinc Corporation (PTC) is a public company, see an income of 31.2 billion in 2005-06, wit an income from operations of Rs. 31.08 against last years 20.32 billions, higher by 53%. Profits after tax are Rs 406.3 millions. It has traded 10.12 billion units as compared to last year's 8.89 billion units. It entered into a long term power purchase agreement with 5 projects totaling 4000 MW, and a cumulative 4600 MW reached. Power trader PTC India saw a total income of Rs. 31.2 billion during 2005-06. The company recorded an income from operations of Rs 31.08 billion as against Rs 20.32 billion in 2004-05, higher by 53 per cent. The profit after tax was lis 406.3 million. The profit before depreciation, interest and tax was higher by 52 per cent at Rs. 619.5 million. The company had trading volumes of 10.12 billion units compared to 8.89 billion units traded in the previous year. During the year, PTC entered into long-term power purchase and sales agree-ments for five projects totaling 4,000 MW capacities. The cumulative capacity tied up through long-term projects by the com-pany accordingly reached 4,600 MW In additions, the company entered into MoDs for another 16 projects totalling 10,400 MW of capacity. Power purchase and sale agreements are being negotiat-ed for these. During the year, the Lanco Amarkantak project in ehhattisgarh, in which PTe had previously picked up an 11 per cent stake, achieved financial clo-sure and entered construction phase. The project has become the first in the country to be granted long-term open access by Power Grid. PTC has also start-ed coal intermediation services and advisory services for the power sector

BHEL: (Bharat Heavy Electrcals Limited)

Bharat Heavy Electrical Limited (BHEL) a power equipment maker in public sector, recorded a net sale of 145 billion in 2005-06, with an increase of 38.5% increase over last year. Net profit is up by 76 % at Rs. 16.8 billion. Power equipments accounted for 72% on total revenue. Its power equipment achieve 402.6 billion units of power in ayear, and recorded 75% PLF. 65 % or a 76741 MW of total installed capacity installation are delivered by the company for India.

Table 12
Turn Over in (Rs. Crores) 2001 to 2006

Utility	*2001-02*	*2002-03*	*2003-04*	*2004-05*	*2005-06*
NTPC	17798	19029	18852	22565	25992.8
NHPC	1221	1172	1414	1581.73	1834
PGCIL	2456	2533	2805	2831.28	3245
NEEPCO	453	496	767	799	834
SJVNL			217	1098.28	1329.59
Total	21928	23230	24055	28875.29	33235.39

Source: MOP Performance Report 2005-06.

Table 13
Net Profits in (Rs. Crores) 2001 to 2006

Utility	*2001-02*	*2002-03*	*2003-04*	*2004-05*	*2005-06*
NTPC	3540	3608	5261*	5807	5706
NHPC	471	510	621	684.56	701
PGCIL	689	643	748	787.8	915
NEEPCO	53	-408	208	196.18	204
SJVNL			-93	282.91	495
Total	4753	4353	6745	7758.47	8021

*Exclude impact of one time dues, settlement of Rs. 3987 Crores.

Source: MOP Performance Report 2005-06.

Net Profit Increases in % of Leading Power Sector Companies 2006-07.

Utility	*2006-07 Net Propfit*	*2005-06 Net Propfit*	*% Change*
NTPC	67264.00	58202.00	15.00
NHPC	5283.20	3847.50	37.32
PGCIL	12290.00	10090.00	21.80
REL	8010.00	65000.00	23.23
TAT Power	6040.70	4717.20	28.06
BHEL	23850.00	16790.00	42.00

Source: Unaudited provisional Results, MOP.

FINANCIAL TURN OVER AND NET PROFIT OF PUBLIC UTILITES

Public utilities turn over and net profit from 2001-02 to 2005-06 show a good performance indicating healthy trend of operational performances. Thermal utilities in private also performed well to have good profit. Budge Budge, Reliance Energy, Dhanu power are classic examples. In 2006-07 REL register a higher % profit.

FINANCIAL PERFORMANCE OF SEB'S

Table 14 shows rate of return 91-92 to 2005-06. This is loss indicator which has increased upto 1999-2000, to -41.5 % then start reducing in 2005-06 to -28.13 %. This is due to refoms effects. ROR: Rate of Return: SEB's losses since 1991-92 increased from -12.7% rate of return (RoR) to -41.5 in 1999-2000. The reforms process of reduction in commercial losses lead to a lowering rate of return (RoR) -37.9 in 2000-01, and -28.12 in 2005-06. The loss making of SEB's start lowering with the effect of reforms and reduction in commercial losses through determined efforts. This is a good beginning.

STATUS OF SEB'S LOSSES (RATE OF RETURN %)

The RoR has negatively increased over the years. The trend value is Yt = -18.82 - 1.51500*t.

Table 14
Status of SEB's Losses

(SEB's Combined Balance Sheet) Financial Performances of State power Sector — *Rs.Crores*

Particulars	*91-92.*	*95-96*	*97-98*	*98-99*	*99-00*	*00-01* RE	*01-02* BE	*02-03*	*03-04*	*04-05* (RE)	*05-06* (AP)
Gross Subsidy on account of Sale to:	7,449	17,282	23,234	30,345	33,814	36,319	41,238	38,836	33,154	34,311	36,002
Agriculture	5,938	13,794	18,296	22,537	24,178	26,301	29,461	26,959	23,346	23,806	25,377
Domestic	1,310	3,158	4,714	7,270	9,174	9,528	11,267	11651	8,885	9,639	10,033
Inter State Sales	201	330	235	538	462	489	510	225	923	866	591
Subventions from State Govts.	2,045	9,229	3,807	7,852	10,938	7,492	8,270	7,980	11,081	9,825	9,831
net Subsidy	5,404	10,053	19,438	22,494	22,876	28,827	32,868	30,855	22,073	24,486	26,170
Surplus generated by Sale to other sectors	2,173	6,660	11,136	6,877	4,242	5,130	5,526	7,499	6,133	6,967	8,640
Uncovered Subsidy	3,231	3,393	8302	15617	18,634 .	23,697	27,342	23,356	15,941	17,520	17,530
Commercial Losses*	4,117	8,324	10,345	18,081	24,920	24,237	28,445	24,320	20,379	20,715	22,013
Rate of Return Without Subsidy (%)	-12.7	-15.1	-17.4	-27.4	-41.5	-37.9	-38.2	-32.08	-28.32	-27.97	-28.13

Source: Planning Commission 2006, * Excluding Subsidy/Subventions.

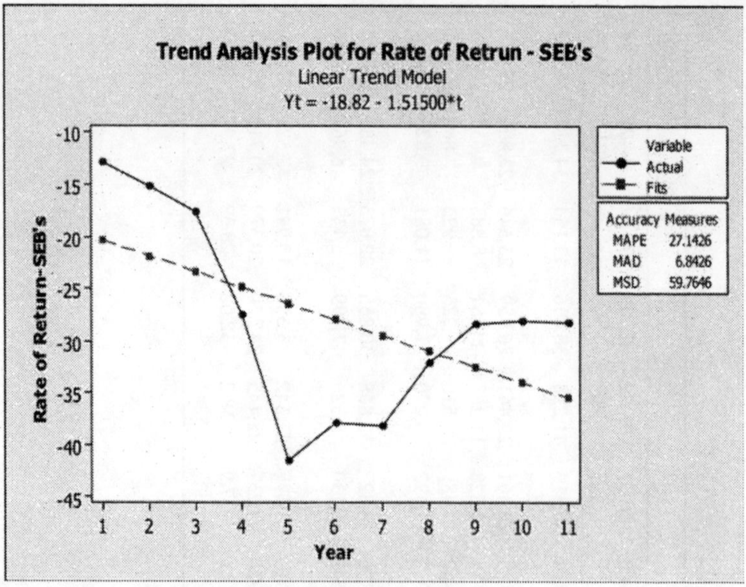

Figure 6

The state utility-SEB's are still incurring losses, but shows an decreasing trend. The combined SEB losses without subsidy declined from Rs. 292 billion in 2001-02 to Rs. 216 billion in 2004-05. Their aggregate turn over grew by 32% during same period to Rs. 1068 billion in 2004-05. Subsidies booked have decreased from 136.7 billions to 110.6 billion.

Of the 70 utilities assessed, 36 showed an improvement in finances, either in terms of an increase in profits or reduction in losses, 29 showed deterioration and five showed no significant change. Forty-three had a positive net worth while 27 had a negative net worth. Similarly, return on capital employed was positive for 44 utilities and negative for 26.

There were other indicators that showed a consistently positive trend. The SEB revenues continued to increase in absolute terms: Rs. 808 billion (2001-02), Rs 882 billion (2002-03), Rs. 981 billion (2003-04) to Rs. 1,068 billion(2004-05).

Table 15
Financial Performance of SEB's

(Rs.Billions)

SEB's Financial Account	*2001-02*	*2002-03*	*2003-04*	*2004-05*
Income (Without Subsidy)-Utilities directly selling to consumers.	808.99	882.43	981.52	1068.81
Expenditure (without tax& depreciation)-Utilities directly selling to consmers	1035.4	1009.48	1090.51	1203.92
Profit/(Loss) (without tax, depreciation and subsidy	226.42	127.05	108.99	135.11
Profit/(Loss) (without tax, and subsidy)	292.52	209.6	195.02	216.67
Subsidy booked	145.95	136.7	104.3	110.16
Book profit/(Loss)	147.36	75.22	92.92	111.1
Subsidy unpaid	51.53	13.3	2.08	7.37
Unrealised revenue	80.88	61.37	42.66	45.04
Profit/(Loss) (without depreciation but on revenue and subsidy realised basis	-189.18	-48.46	-22.68	-34.38

Source: PFC Report 2004-05.

Growth in the sale of power in million kWh was 7.43 per cent in 2004-05 where-as it was 7.86 per cent in 2003-04. Further, growth in revenue from sale of energy in percentage terms declined marginally from 10.64 in 2003-04 to 9.42 per cent in 2004-05.

Category wise, agricultural revenue increased by 8.5 per cent while industrial revenue rose by 9.9 per cent. The share of Agriculture and industry in the total rev-enue increased marginally from 6.67 per cent to 6.72 per cent and from 47.22 per cent to 48.27 per cent respectively.

A cause for worry was rising subsidies. These increased from Rs 104 billion in 2003-04 to Rs 110 billion in 2004-05. The level of subsidy released by the state governments also increased from Rs 10.63 billion in 2003-04 to Rs 11.75 billion in 2004-05. The Chhattisgarh, Karnataka, Kerala and Rajasthan governments did release the full subsidy during 2004-05.

FINANCIAL PERFORMANCE OF STATE POWER SECTOR

The rate of return of the state sector which was (negative)–24.01 % in 2006-07 improved marginally to -18 % (without subsidy) in the revised estimate for 2007-08 as per planning Commission estimates.

NTPC'S FINANCIAL PERFORMANCE BEFORE AND AFTER REFORMS YEAR WISE

Thermal Power utility NTPC which generate nation 31% of power, has set a record trend of running with high profit before and after reforms with stern determinations and dedication in the field.

NTPC premier thermal utility generates nation's more than 31% of power, have profits before reforms, up to 1990 are marginal, same utility after improving operational performances increased profits on higher side from 1990-91 to 2005-06. In the year 2003-04 it has one time settlement of SEB's dues payment received Rs. 3987 croes. Performance of NTPC is exemplary.

Table 16
NTPC's Turnover and Profits Before Reforms

Year	*Turnover (Crores)*	*Net Profit (Crores)*
1983-84	145.7	45.96
1984-85	333.6	87.54
1985-86	522.4	183.04
1986-87	637.4	183.04
1987-88	847	211.86
1988-89	1263.2	302.4
1989-90	2036.2	380.82

Source: NTPC Annual Reports 1982-1990.

Table 17
NTPC's Turnover and Profits After Reforms

Year	*Turnover Rs. (Crores)*	*Net Profit- Rs. (Crores)*
1990-91	2400.6	536.6
2001-02	17798	3540
2002-03	19029	3608
2003-04	18852	5261*
2004-05	22565	5807
2005-06	25992.8	5706

*Exclude impact of one time dues, settlement of Rs. 3987 Crores.

Source: NTPC Annual Reports 1982-1990.

TARIFF

Electricity supply services are the result of combination of various technologies, infrastructures that require large investments, labor and technical know-how material and primary energy sources. Each of these carries a price tag and they are partly substitutable for one another. They are generation, transmission and distribution and their cost form total cost of power. Tariff is the price fixed by power seller or distributor to end user the customer.

ENERGY CHAIN

Various sources are of power are extracted from primary energy source. These sources are treated and converted by technologies as power in power plants or mills for generation of power. Generated power is transmitted by its technology. Finally power is distributed to user, measured by meters and billed according to meter measurement, by tariff applicable to the consumer.

REGULATORY COMMISSIONS

Tariff of state SEB's are the main causes of huge losses that dose not meet cost of power supplied to consumer. Power sector reforms take care of tariff through Regulatory commissions in State and Central. They are constituted under Electricity Regulatory Commission Act 1998 and Electricity Act 2003. They consist a Chairman and Members. Their functions are to regulate tariff of generating companies in state and central as the case may be.

- To determine tariff of electricity, tariff orders to determine tariff on transmission, distribution,
- To issue licenses to distribute power on competitive basis.
- To specify and enforce with respect to quality, continuity and reliability of services by licensees.
- To fix trading margins and specify grid codes and disciplines.
- To play proactive role for open access to consumers and for harnessing captive generation capacities.
- To adjudicate disputes related to generation, transmission and distribution in relation to codes, disciplines, tariffs, trading of power, consumer redressal, licensing of transmission, distribution matters in respect of their state and central spheres, etc.

Electricity tariffs (as per latest tariff order; maximum tariff charged in a category) as per regulatory commission orders.

Various state regulatory commissions tariff orders in domestic, HT and agriculture are given shows A.P and W.Bengal states higher tariff in domestic consumers.

W. Bengal charges higher HT tariff. Assam Charges higher agriculture tariff. Latest tariff orders of various states show tariff changes.

Table 18

Various State Regulatory Commissions Latest Tariff Orders are given below Consumer Wise

	Domestic		*HT*		*Agriculture*	
SEB	*Fixed Charges (Rs/month)*	*Energy Charges (Rs/Unit)*	*Fixed Charges (Rs/month)*	*Energy Charges (Rs/Unit)*	*Fixed Charges (Rs/month)*	*Energy Charges (Rs/month)*
A Pradesh	—	5.50	195/kVA	4.40	44	—
Assam	30	4.30	140/kVA	4.00	40/kVA	3.15
Bihar	24	1.60	125/kVA	1.78	125/BHP	1.78
Chhattisgarh	100	2.05	310/kVA	3.80	65/HP	1.00
Delhi	—	—	150/kVA	4.30	12/kW	1.50
Goa	—	2.50	150/kVA	3.00	—	1.00
Gujarat	15	4.70	215/kVA	4.10	10/BHP	0.50
Haryana	—	4.25	—	4.09	100/BHP	0.62
Himachal Pradesh	—	2.75	125/kVA	2.20	70/kW	2.35
Jammu & Kashmir	—	1.15	—	—	—	—
Jharkhand	40/connection	1.70	140/kVA	4.00	200/HP	0.75
Karnataka	50kW	4.70	180/kVA	4.30	20/HP	0.40
Kerala	—	5.25	270/kVA	3.00	6/kW	0.65
Madhya Pradesh	475/connectin#	3.40	400/kVA	4.10	—	2.50
Maharashtra##	40kVA	4.00	—	2.10	180/HP	1.10

Contd...

SEB	*Domestic* Fixed Charges (Rs/month)	*Domestic* Energy Charges (Rs/Unit)	*HT* Fixed Charges (Rs/month)	*HT* Energy Charges (Rs/Unit)	*Agriculture* Fixed Charges (Rs/month)	*Agriculture* Energy Charges (Rs/month)
Meghalaya	—	3.20	240/kVA	3.20	—	1.20
Orissa	—	4.50	200/kVA	3.00	30/kW	1.10
Punjab	—	3.89	—	3.72	208/BHP	—
Rajasthan	105	3.50	60/kVA	4.01	50	230/HP*
Tamil Nadu	5	3.05	300/kVA	3.50	21/HP	2.00
U.P	51kW	3.00	180/kVA	3.50	15/BHP	0.75
Uttaranchal	—	2.00	350/kVA	1.90	105/BHP	0.70
W. Bengal	—	5.77	180/kVA	5.55	—	1.95

Source : Tariff orders

120 free units allowed; ## REL/TEL metered tariff for agricultural consumers : FCC Rs.374 per kVA, EC Rs.2.85 per unit; * Per month SOurce: SERC 2006

Table 19
Previous Tariff Orders (Maximum Tariff Charged in a Category) As Per Regulatory Commission Orders

	Domestic		*HT*		*Agriculture*	
SEB	*Fixed Charges (Rs/month)*	*Energy Charges (Rs/Unit)*	*Fixed Charges (Rs/month)*	*Energy Charges (Rs/Unit)*	*Fixed Charges (Rs/month)*	*Energy Charges (Rs/month)*
Andhra Pradesh	—	5.50	36/HP	4.40	48/HP	—
Assam	—	4.00	10/kW	4.70	—	3.35
Bihar	24	1.60	125/kVA	1.78	125/BHP	1.78
Chhattisgarh	100	2.05	—	4.42	—	—
Delhi	10/kW	3.85	—	—	—	1.10
Goa	—	2.50	150/kVA	3.00	—	1.00
Gujarat	15	4.70	205/kVA	4.10	10/BHP	0.50
Haryana	—	4.25	—	4.09	100/BHP	0.62
Himachal Pradesh	—	2.40	125/kVA	1.90	70/kW	2.35
Jammu & Kashmir	—	1.15	—	—	—	—
Jharkhand	65 or 1.00/Unit	1.70	140/kVA	4.00	200/HP	0.75
Karnataka	0.75/kVa	2.85	180/kVA	4.30	67/HP	1.75
Kerala	170	5.45	350/kVA	3.20	—	—
Madhya Pradesh	150	2.85	150/kVA	3.87	80/HP	00
Maharashtra##	100/Connection	4.00	60/HP	2.50	180/HP	1.10
Meghalaya	—	2.50	120/kVA	2.00	—	0.90

Contd...

SEB	Domestic		HT		Agriculture	
	Fixed Charges (Rs/month)	Energy Charges (Rs/Unit)	Fixed Charges (Rs/month)	Energy Charges (Rs/Unit)	Fixed Charges (Rs/month)	Energy Charges (Rs/month)
Orissa	30/kW	3.20	200/kVA	2.90	30/kW	—
Punjab	—	3.64	—	3.66	205/BHP*	2.00*
Rajasthan	75	2.75	60/kVA	4.01	85/HP	1.65/HP
Tamil Nadu	5	4.75	300/kVA	3.50	21/HP	2.00
Uttar Pradesh	250/unit	3.20	180/kVA	3.75	70/BHP	10/BHP/month
Uttaranchal	—	1.80	125/kVA	1.90	50/BHP	0.70
West Bengal	—	5.00	180/kVA	3.30	—	1.83

Source: SERC Reports.

* Without subsidy (with subsidy – EC: Re.0.57 per unit and FC: Rs. 60 per BHP), *Source*: 2006.

Various previous state regulatory commissions tariff orders in domestic, HT and in agriculture are given. The commission orders indicate tariff randes in states and help rationalize tariff.

Table 20
Breakup of Average
Unit Cost of Electricity Sold in 1990-91 to 2000-01

Component	*Unit cost of sale in paise per Kwh 1990-91*	*Unit cost of sale in paise per Kwh 2000-01*
Cost of power supplied to grid A+B	83.8	203.9
A.Fixed cost	20	34
Depreciation	4.9	12.3
Interest	15.1	21.7
B.Variable Cost	63.8	169.9
Fuel	20.1	28.9
Power purchase include state Excise Duty	21.6	103
Operation & Maintenace cost	4.2	5.5
Employee cost	14.2	27.6
Misc expenses- Ex. Duty on aux. consumption	3.7	4.9
II. Cost of power loss in transmission (at 6%)	5.4	13
Cost of power supplied to distribution section	89.2	216.9
III. Cost of power lost in distribution	19.5	87
Cost of power upplied to consumer (I+II+II)	108.7	303.9

Source: Planning Comissisio Annual Report on the workinf of EB's & ED & MOP 2000-01.

Theses are based on T&D loss of 22.9% and 32.9% during 1990-01 and 2000-01 respectively. Average unit cost of power in 1990-91 is 108.7 paise and in 200-01 are 303.9. The cost includes various components of power and fetches revenues to SEB's on tariff fixed and in turn paid to thermal generators and utilities.

Table 21
State-wise Average
Unit Cost and Tariff of Power Sold in 1975-76 and 2000-01

	Cost of power in paisa per Kwh		*Average tariff in pasia per Kwh*		*Percentage increase in 2000-01 over 1975-76*	
SEB	*75-76*	*00-01*	*75-76*	*00-01*	*Cost*	*Tariff*
A. Pradesh	30.27	343.35	30.06	211	1034.43	601.9
Assam	28.9	556.75	25.37	351.24	1826.5	1284.5
Bihar	31.86	391.79	26.09	251.8	1129.7	865.1
Delhi	n.a.	474.69	n.a.	296.9	n.a.	n.a
Gujarat	25.91	354.47	23.85	215	1268.1	801.5
Haryana	22.7	403.7	19.35	205.36	1678.4	961.3
Himachal	31.7	238.3	21.65	220.9	651.7	920.3
J&K	n.a.	432.34	n.a.	137.03	n.a.	n.a.
Karnatka	31.69	324.56	14.57	217.2	2270.8	1390.7
Kerala	16.52	303.05	13.3	189.13	1734.4	1322
MP	23.11	314.41	21.36	204.92	1260.5	859.4
Maharastra	20.14	318.69	19.91	271	1482.2	1261.1
Meghalaya	25.15	236.03	12.58	161.02	838.5	1180
Orissa	16.68	177.16	14.1	272.41	962.1	1832
Punjab	20.58	274.53	15.77	177.29	1234	1024.2
Rajasthan	22.79	270.77	25.01	224	984.4	795.6
TamilNadu	24.97	270.77	25.01	224	984.4	795.6
UP	34.76	361.06	26/08	258.57	938.7	891.5
W. Bengal	25.67	344.3	24.63	223.66	1241.2	808.1

Source: Planning Commission Annual Report on the Working of SEBs and EDs – various issues.

The cost of power and tariff/sold price of power in each state for 1975-76 and 2000-01 show wide variation. Cost of power has a variation of 1034.43 % to a maximum of 2270 %. Tariff/ Sold price variation are 601.9 % to a maximum of 1832 %. From this it indicates that the tariff and cost of power have increased many folds. But expect state of Orissa no state have sold or fixed average tariff more than cost of power. This is leading to huge losses. In turn this loss will make burden to the thermal power generators who supply more than 80% of power to the nation.

Table 22

Year	*Average Cost of Supply (ps/k/Wh)*	*Average Traffic (ps/k/Wh)*	*Shortfall (losses) (ps/k/Wh)*	*Recovery through Tariff%*
1992-93	128.2	105.4	22.8	82.2
1993-94	149.1	116.7	32.4	78.3
1994.95	163.4	128.0	35.4	78.3
1995-96	179.6	139.0	40.6	77.4
1996-97	215.6	165.3	50.3	76.7
1997-98	239.7	180.3	59.4	75.2
1998-99	263.1	186.8	76.3	71.0
1999-2000	305.1	207	98.1	67.8
2000-01 (RE)	327.3	226.3	101	69.1
2001-02 (AP)	349.9	239.9	110	68.6

Source: Planning Commission-Year-2002.

COST OF SUPPLY VERSUS TARIFF

The average cost of supply and average tariff has a gap or short fall.

The recoveries of cost through tariff or sold price at 1992-93 are 82.2%, but in 2001-02 the recovery is 68.6 only. This will again lead to a huge loss to power sellers. Theses recoveries are manageable only by power commercial loss reduction methods to make healthy of financial aspects, that will be affecting thermal power generations also.

Table 23

Year	*Average Real Cost of Supply (ps/k/Wh)*	*Average Real Traffic (ps/k/Wh)*
1992-93	55.02	42.84
1993-94	61.1	41.92
1994.95	149.9	117.43
1995-96	158.93	120.86
1996-97	184.02	127.15
1997-98	208.43	121.82
1998-99	230.78	122.90
1999-2000	263.01	107.25
2000-01	257.48	119.41
2001-02	269.15	103.85

Source: Planning Commission 2003.

COST OF SUPPLY VERSUS TARIFF AND REAL VALUE

The real value of average cost of supply and tariff these are indicated above, year wise from 1992-93 to 2001-02.The real values have a change of 55.02 in 1992-93 to in 2001-02 for 269.15 in cost of supply.

Year wise from 1992-93 to 2001-02.The real values of tariff have a change of 42.84 in 1992-93 to in 2001-02 for 103.85. This has comparison with actual average cost of supply and average tariff.

AVERAGE COST TO AVERAGE REVENUE

The Average Cost of Supply (ACS) and Average Revenue Realized (ARR) gap is an important measure per unit of power. The gap being analysed.

Table 24
Average Cost to Average Revenue Gap

Year	*ACS*	*ARR*	*Gap*
1996-97	1.62	1.24	0.38
1997-98	1.82	1.36	0.46
1998-99	1.96	1.39	0.57
1999-2000	2.12	1.44	0.68
2000-01	2.3	1.59	0.71
2001-02	2.46	1.81	0.65
2002-03	2.39	1.96	0.43
2003-04	2.4	2.01	0.39
2004-05	2.5	2.3	0.2

Source: PFC Report 2004-05.

The Average Cost of Supply (ACS) increased from 1.62 in 1996-97, to Rs. 2.50 per unit in 2004-05 an increase of 54.32 %. The Average Revenue Realized (ARR) increased from Rs. 1.24 per unit to Rs. 2.30 per unit. The gap between ACS and ARR decreased from 0.38 to 0.20 per unit during this period. The gap if it widen will increase more losses.

Reducing gap is a good indicator of loss reductions. However this is dependent on commercial losses, and its reductions.

YEAR WISE COMMERCIAL LOSSES OF SEB's

Pre and post reforms commercial losses of SEB's are given below:

Table 25

Annual Commercial Losses of SEBs (Rupees billion)

	Subsidy given by SEBs to		*Cross subsidy collected from*	*Net subsidy*	*Amount Subvention*	*Losses without*	*Losses after*
Year	*Agricultural consumers*	*Domestic consumers*	*Industrial & commercial consumers*	*Provision by SEBs*	*Received from state government*	*Receipt of subvention*	*Receipt of subvention*
1985-86	n.a.	n.a.	n.a.	n.a.	n.a.	15.2	n.a.
1987-88	n.a.	n.a.	n.a.	n.a.	n.a.	23.1	n.a.
1989-90	n.a.	n.a.	n.a.	n.a.	n.a.	40.2	n.a.
1990-91	46.05	12.87	12.96	45.96	14.97	30.8	15.8
1991-92	58.59	14.82	28.09	45.32	12.02	40.2	28.2
1992-93	73.35	20.35	39.11	54.59	31.82	45.6	13.8
1993-94	89.66	21.31	45.23	65.74	23.54	50.6	27.1
1994-95	109.41	25.37	53.79	80.99	51.27	61.3	10
1995-96	136.06	32.24	63.33	104.97	75.92	87.7	11.8
1996-97	155.86	43.86	77.75	121.97	66.31	113.1	46.8
1997-98	190.21	52.58	90.11	152.68	63.65	139.6	76
1998-99	224.73	63.32	83.1	204.95	103.52	208.6	105.1
1999-2000	246.5	81.21	52.64	274.67	112.65	263.5	150.8
2000-01	269.5	100.4	57.91	3111.95	74.65	291.3	216.6

Source: Planning Commission: Annual Report on the Working of SEBs June, 2002 and earlier issues.

ANNUAL COMMERCIAL LOSSES OF SEBS

Commercial losses in pre and post reforms from 1985-86 to 2000-01 are shown in Table 25, the commercial losses are 15.2 billion in 1985 and rose to 216 billion rupees. The raise in these losses are due to huge theft of power ranging on an average more than 20% to 60% . This is more in north east region. Pilferage or thefts are the main causes of huge commercial losses. Reduction efforts through high voltage distribution system, theft deductions, accurate metering is programmed to reduce theft of power as primary activity.

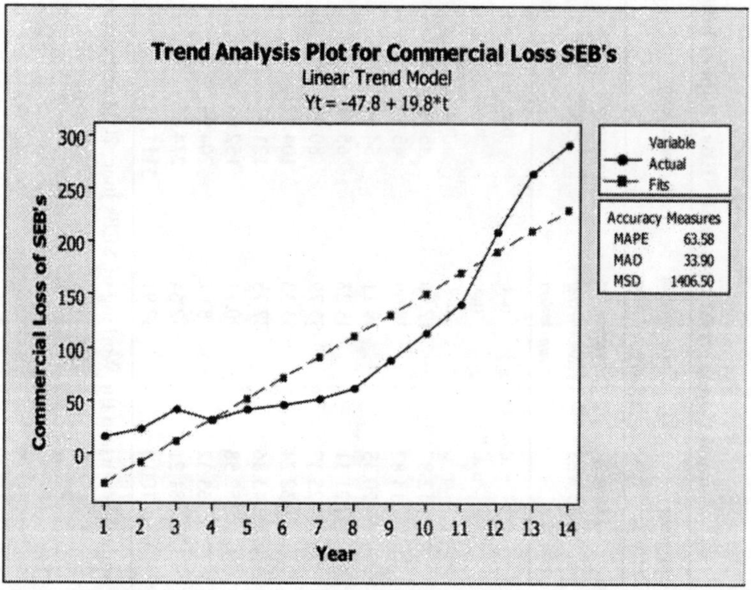

Figure 7

COMMERCIAL LOSSES OF DISTRIBUTORS

Commercial Losses of power Distributors (State/Utilities)

AT & C Losses for 2002-03,2003-04 and 2004-05 for state and utilities are given in Percentage.

Table 26

%

State/utility	*2002-03*	*2003-04*	*2004-05*
APCPDCL	29.68	19.12	23.95*
APEPDCL	19.32	17.79	14.27
APNPDCL	26.84	10.41*	21.91
APSPDCL	27.26	17.17	20.55
Arunachal Pradesh	61.80	16.26#	37.26*
Assam	39.44	43.35	40.40
Bihar	77.64	77.00	74.09
Chhattisgarh	38.68	38.87	45.07
BSES Rajdhani (Delhi)	47.40	45.07	40.64
BSES Yamuna (Delhi)	61.89	55.29	50.12
NDPL (Delhi)	47.79	44.86	33.79
Goa	22.98	21.28	17.27
Gujarat	31.24	30.98	30.32
DHBVNL (Haryana)	41.39	48.68	36.39
UHBVNL (Haryana)	42.82	52.84	43.12
Himachal Pradesh	29.52	9.27*	21.71*
Jammu & Kashmir	61.71	67.56	68.33
Jharkhand	53.00	65.00	69.23
Bescom	35.70	28.91	24.83
Gescom	43.54	43.86	42.99
Hescom	47.73	31.66	41.64
Mescom	35.68	25.82	26.62
Kerala	29.08	27.44	32.12
Poorva KVVCL	34.05	42.68	44.37
Paschim KVVCL	43.41	45.66	38.86
Madhya KVVCL	51.13	57.06	53.17
Maharashtra	44.30	44.18	26.62*
Manipur	76.88	69.73	88.50*
Meghalaya	42.34	39.31	38.16
Mizoram	49.55	40.75	22.11*

Contd...

%

State/ utility	*2002-03*	*2003-04*	*2004-05*
Nagaland	53.82	55.59	39.53*
Cesco	55.04	51.10	49.37
Nesco	52.25	50.36	42.96
Sesco	49.76	51.56	45.71
Wesco	47.30	46.18	40.60
Pondicherry	41.67	41.01	40.49
Punjab	23.19	22.40	20.62
AVVNL	41.22	44.57	49.75
Jd.VVNL	41.97	41.56	47.56

Source: MOP 2006, * official figures # includes power trading.

Commercial losses at transmission, distribution and pilferage/ theft losses in 2002-03 to 2004-05 are given as above. Power distribution un-bundling programme bring new utilities, instead of SEB's activity. Table 26 show utilities loss % in years. National average of commercial losses are 38.6% in 2002-03. In 2003-04 are 37.75% and in 2005-06 are 36.77% indicating a slow reduction in AT&C losses. The losses reduction by APDRP programme, theft prevention, billing collection, metering are vigorously undertaken to reduce commercial losses. APEPDCL, Tamil Nadu Goa, CESE have shown low % of losses as compared to Bihar, J&K, Manipur, Sikkim, Jarkhand whose higher losses attribute to more pilferage and theft of power.

STATE RANKING

In order to create healthy competition among states and to reward the states during power sector reforms period a sectoral rating is evolved among states and carried out through out side professional agencies, organized by MOP. Two rating agencies–ICRA and CRISIL–had done this excellent work of rating. This has been a tedious exercise, marked by all kinds of methods, including arguments and discussions with the state's on the criteria used, or struggles to get updated and authentic information at various utility offices. The drafted report was processed by the ministry for over two months before the ranking announced. The ranking is based on data collected upto December 2005. A common assessment model was drafted. This study was following previous year method. It assessed commercial viability of the power sector in the states, the progress in reforms, sustainability of the business model adopted and the facilitation of a competitive climate.

The work made a comprehensive review of the states on eight basic parameters. They are state government-related parameters; regulatory process, business risk analysis, financial risk analysis, other–MIS, etc., progress in attaining commercial viability, sustainability of the revenue model, and creation of a competitive climate.

Parameters–(Table 27) of last two–introduced, for "forward looking" and sustainability of state's approach. These parameter had a weightage of 12.5 per cent. Weightage of the other six parameters, that are historical were freezed to 75 per cent.

In case more than 60% of the states had achieved compliance on a certain parameter, the states that complied got zero score. Those that did not, were given negative marks. More over, negative marks were assigned for states which offered free power. Norms like PLF were freezed.

STAR SCORERS

Top scorer was Delhi on the state government parameter, with the government maintaining excellent assistance for reforming the state.

Coming to regulatory processes parameter, Andhra Pradesh maintain top, as the state regulator making an excellent work.

In T&D, the state of Himachal Pradesh become first, overtaking Goa to number two position.

The state of Andhra Pradesh maintain first position in financial risk assessment. The state of Meghalaya has overtaken Karnataka to get the first position in generation, relegating Karnataka to number two.

The state of Goa top scorered in sustainability of its revenue model, recording consistent profits from 2002-03.

The state of Goa made it to top in commercial viability by good cost coverage (by trading help) and maintained.low T&D losses. The state of Gujarat scored high through a competitive environment by unbundling, intrastate ABT implementation, and durable captive power Policy.

NEGATIVE MARKS

The states of Goa, Punjab, Meghalaya, Jharkhand, Bihar and Jammu & Kashmir scored negative marks for lack of initiative. They got penalty marks for their subsidies and free power, and delay in the area of reforms and restructuring.

The states of Assam, Haryana, Madhya Pradesh, Bihar and Manipur got negative marks for the position of non-commercial viability for their financial ill-health, more dependence on subsides and lower cost coverage.

In the regulatory process, the states of Goa, Punjab, Meghalaya, Sikkim, Mizoram, Arunachal Pradesh, Nagaland, Bihar and Manipur scored negative marks for not creating SERCs and for their remaining non-functional or for non- passing of orders.

The states of Jharkhand, Arunachal Pradesh, Bihar, Manipur and Jammu & Kashmir scored negative marks in high T&D losses and poor networks.

IMPROVEMENT IN CASH COLLECTIONS

Due to distribution reforms, anti theft measures, separated agri-feeders, cash collection, excluding subsidy, have improved, tariff increased in many states and revenue were increased by trading in power.

DECLINING TREND IN ADJUSTED BOOK LOSSES

Indicate the financial health of the utilities are adjusted book losses, which measure profit or loss after excluding subsidy support and factoring in the build-up of receivables. Declining trend are shown by adjusted book losses, though substantial in absolute terms. Indications of payment for power purchase and fuel is developing and the utilities' debt servicing capability.

DEPENDENCE ON SUBSIDY

Bigger states, Andhra Pradesh, Gujarat, Karnataka, Rajasthan, Haryana, Punjab and Maharashtra continue to have much of dependence (as mush as 25% of revenues) on state subsidy, despite sudsidy reduction take place. In some states subsidies are in increasing trend as in Rajasthan, Haryana, Punjab and Maharashtra. Famous declaration on free power will ruin and enhance the situation.

MORE GAPS IN ARR AND ACS, WITHOUT SUBSIDY

More gap persist in Delhi Punjab, Karnatak, Rajasthan, Uttar Pradesh, Madhya Pradesh and Haryana. The gap is reducing in many states like Gujarat, Delhi, West Bengal, Maharashtra, Kerala, Assam, Tripura, Meghalaya and Uttaranchal which is a good sign.

ACS IN CHECK

Improvements in efficiency is a good sign, renegotiation of high-cost PPAs and decline in interest costs, recorded by states of Andhra Pradesh, Gujarat,Delhi, Karnataka and west Bengal. They managed to contain the average cost of supply, even at the stages of fuel prices increase.

AT&C LOSSES STAND MODERATE-TO-VERY HIGH

AT&C losses are pathetic. Many states, including Delhi, have AT&C loses at 30-40 per cent.

Ratio of billed energy on metered basis shows low, despite progress made in consumer metering, the proportion of energy billed on a metered basis remains low. Unmetered consumption are huge form agriculture front. They are the prime loss makers.

POWER TRADING

North-eastern states have been able to earn precious UI charges and improve their financial viability, though the rating agencies doubt the sustainability of these charges, during this trading boom. And with power surplus stages. Cost coverage, are not adequate in north-east.

UNBUNDLING

The states of Gujarat, Maharashtra, Assam and Madhya Prades had restructured recently. States like Kerala, Tamil Nadu, Punjab, Bihar and Chhattisgarh not completed their restructuring. States that have unbundled, the discoms are limited functional and financial autonomy, they are at distant position.

Competitive Markets: Provisions of open access and trading will come fully into play and make the power market more competitive and more stronger after a period.

MIS AND DATA

Reliable and updated data are difficult to get from the state utilities. Restructuring of SEBs has helped to get data, they should be recorded from utilities of a state. This important work should be systamatised.

The power sector ratings are a good and healthy competition, among the states and compete them more enthusiastically. Each parameters are measured and assessed, the states meet each other in a competitive environment to out do. The scores form a state of comparison, making other states to gauge progress and make a healthy competition.

The exercise are executed by reputed rating agencies; with scientific methods with open-ness. The study forms a basis for providing incentives under the APDRP and for the central budgetary assistance. This has become a periodical evaluation method on a national level involving all states.

STATE AT&C LOSSES

Group-wise Technical and Commercial Losses of States are are given as under.

Table 27

State At & C Losses

Less than 20 per cent	*20-30 per cent*	*30-40 per cent*	*More than 40 per cent*
Tamil Nadu, Goa	Andhra Pradesh, Gujarat, West Bengal, Himachal Pradesh, Maharashtra, Tripura, Uttaranchal, Punjab	Karnataka, Kerala, Sikkim, Meghalaya, Chhattisgarh, Mizoram	Delhi, U.P. Assam, M.P., Arunachal Pradesh, Manipur, Bihar, Rajasthan, Haryana, Jharkhan, Nagaland

Source: MOP-2006.

State's AT & C losses less than 20%, 20-30%, 30-40%, More than 40% are shown above. They indicate the status of power losses, to make reduction efforts. States at higher losses have to evolve strategies to reduce power losses. The groupings make their status for improvement and remedies.

REFORMS AND RESTRUCTURING–KEY HIGHLIGHTS:

Status of reforms and restructure of various states are given below:

The status report of states reforms compliance show that still many states have not-bundled- (13 bundled out of 37). Their electricity boards reform insists this unbundling as a first step to be carried out by SEB's for effectiveness of working. North eastern states are yet to take up reforms in a serious manner. The unbundling of SEB's splits generation, transmission and distribution into separate activity to specialize systems for effective performances. Table 28 indicate SEB's casual reform approach, slow pace.

Table 28

State	Constitution of SERC	No. of tariff orders issued	Year of last tariff order	Unbundling / corporatisation (nos)	APDRP sanctions (Rs billion)	APDRP releases (Rs billion)
Delhi	1999	4	2005	2002 (3)	9.23	1.06
Haryana	1998	4	2005	1999 (2)	4.32	1.69
Him. Pradesh	2001	4	2006	Ext Dec. 2006	3.23	2.42
J & Kashmir	2004	X	X	X	11	4.09
Punjab	1999	4	2005	Ext Dec. 2006	7.16	1.79
Rajasthan	1999	2	2004	2000 (3)	11.93	3.86
Uttar Pradesh	1998	5	2004	2000 (4)	10.91	2.37
Uttaranchal	2002	2	2005	2001 (1)	3.1	2.8
Chhattisgarh	2001	1	2005	Ext Dec. 2006	4.08	1.59
Gujarat	1998	2	2004	2005 (4)	10.8	4
Goa	2002	X	X	X	2.94	1.13
Madhya Pradesh	1998	5	2006	2002 (3)	6.63	1.3
Maharashtra	1999	4	2004	2005	22.3	3.74
Andhra Pradesh	1999	7	2006	1999 (4)	14.6	5.67
Karnataka	1999	5	2005	1999 (5)	11.9	4.48
Kerala	2002	3	2005	Ext Dec. 2006	8.64	2.31
Tamil Nadu	1999	1	2003	Ext Dec. 2006	9.48	4.42
Bihar	2005	X	X	Ext Sept. 2006	8.54	3.13
Jharkhan	2003	1	2004	Ext Oct. 2006	4.24	1.54

Contd...

State	*Constitution of SERC*	*No. of tariff orders issued*	*Year of last tariff order*	*Unbundling / corporatisation (nos)*	*APDRP sanctions (Rs billion)*	*APDRP releases (Rs billion)*
Orissa	1996	6	2005	1996 (4)	5.92	0.74
W. Bengal	1999	4	2006	Ext Sept. 2006	4.42	0.93
Assam	2001	3	2005	2004 (3)	6.51	3.49
Meghalaya	2006	X	X	Ext Dec. 2006	2.27	0.58
Arunachal Pradesh	X	X	X	X	0.8	0.4
Mizoram	JERC notified	X	X	X	1.09	0.78
Manipur	JERC notified	X	X	X	1.4	0.03
Nagaland	X	X	X	X	1.2	0.69
Tripura	2004	1	2005	2005 (1)	1.47	0.54
Sikkim	2005	X	X	X	1.64	1.55
Total	25		20	13 (37)	191.8	63.07

Source: MOP 2006.

SECTION 123 OF US ATOMIC ENERGY ACT, US AND INDIA BILATERAL NUCLEAR COOPERATION AGREEMENT UNDER CONSIDERATION AND DISCUSSION

Share of nuclear power in total installed capacity in India as on 2007 are 2.95% over 1,32, 329 MW as per Rajya sabha statistics on question No. 420.

123 agreement give three major advantages, (1) Global civilian nuclear cooperation. (2) Right to retain nuclear weapons while being outside the Non-Proliferation Treaty (NTP). (3) Access to high and dual–use technology. By this US will help India (a) uninterrupted supply and reserve fuel of nuclear fuel to nuclear reactors. (b) In return India will put its agreed civilian nuclear reactors in perpetuity under IAEA. (c) If India does nuclear tests, the agreement will be terminated with one years notice by the US and will have right to return of all equipments and materials transferred. (d) While outside NPT, India will be allowed to have nuclear weapon. (e) India get resource to meet 20,000 MW capacity by 2020. On concerns the transfer of fuel and components are governed by US laws and not 123 agreement. Uncertainty over the eventual shape of national reprocessing facilities under IAEA safeguards. The likely-hood of US interference in India's national security/foreign policy issues and indirectly hints to align with US policy and rocked our parliament for no-confidence motion.

SUSTAINABLE POWER DEVELOPMENT

Effective pollution control measures, disposal of large quantities of ash from coal based thermal power utilities, environmental care, are explained in Blue Print for Power Sector Development-2001 and Annual Report 2005-06, Ministry of Power, Government of India, that are given below.

NTPC and other PSU power utilities jointly set up a Special Purpose Vehicle, (SPV) registered as a society to take up afforestation and environment measures to reduce carbon-dioxide in the atmosphere. The objectives are:

Fruitful channelising investments by members to increase national forest cover. To identify suitable afforestation lands with Ministry of Environments and Forests (MOEF) that co-ordinate with state's authorities and to facilitate quick forest clearance and procurement for future projects, for NTPC and its members. To make compensation for forest land take over and afforestation expenditures through MOEF for the diversion of forest land.

Under fly Ash Utilisation plan, 90 million tons of fly ash per annum are estimated as on 2001 of all coal based power utilities. They are directed to be utilized as per "The Fly Ash Mission" (TIFAC) had made useful recommendations. Utilise fly ash in the manufacture of cements bricks, pavement materials and blocks, floor tiles, wall panels, road construction, land filling and back filling of mines. So that they are used with cost benefit in major constructions by the findings of laboratories as in developed counties and to contain damage control with gains.

Services of (TERI) TATA Energy Research Institute have been engaged for clean Development mechanism (CDM) under improvement in environmental performance for project formulation base line survey, developed countries practice tie-up, Corbon-dioxide emission reduction and supervision of project implementations.

Noise pollutions are controlled by silencers, noise absorbents, noise barriers, noise proof control rooms, green belt around plant that reduces noise levels with construction regulations and noise pollution approvals in all plants.

Under Clean coal technology, that reduces NOX, SOX, CO2 emissions in coal fired units. Not using high ash (higher than 34%) level coal mix are notified by MOEF with effect from Ist June 2001 in all power plants.

All Thermal power utilities are given approval for running them in the recent years, only with inbuilt pollution control measures, that takes care of safe limits of emission of NOX Nitrogen oxide, SOX sulphur dioxide, CO2 corbodioxide, O3 ozone and smoke which are let out beyond safe height as per

CEA and Central Pollution Board norms. So that public atmosphere are protected and safe-guarded. Old plants which do not comply with this are given time to impart pollution control measures. Otherwise their operations are not permitted as in the case of Reliance Power's, Dhanu Thermal, Dhanu in Maharastra to comply with pollution board norms.

Water balance improvement and norms are fixed for all power plants for its implementation. "Renovation and Modernisation" scheme after post reform period that take care of environmental aspects also for old plants in a strict manner, so that old power plants not adhering norms are brought under control. Even sick State Electricity Board utilities are availing financial assistance for this purpose to fall in line.

(NEERI) National Environmental Engineering Research Institute, Nagpur, monitors and offer suggestions for short and long term measures for Thermal and Hydel power utilities since 2001.

Continuous research and development are undertaken with the direction of Ministry of Power in leading national laboratories, to improve and develop various environmental aspects of power plants for its implementations.

ENVIRONMENTAL PERFORMANCES OF THERMAL UTILITIES

NTPC that generate nation's more than 40% of power had achieved ISO-14001 Standards for certified environment management system in its all power plants. NTPC has taken fool proof measures in pollution and environmental damage in the last decade. Other utilities like NEEPCO in North East India, others are following this as an example to acquire it.

NTPC had planted over 1.45 crore trees. Its effort had brought down to 30'C around Ramagundam Power Station due to afforestation as revaled by a study by National Remote Sensing Agency (NRSA), Hyderabad.

Heat-rate improvement by State Electricity Boards and NTPC plants had made a big headway. Dhanu Thermal, Bridge-Bridge in W.B, in private sector and other state sector plants had demonstrated their superiority in heat-rate improvement. Super critical technology are introduced in Indian Power Sector through NTPC plants and later for super critical boiler technology introduction. Tie up with USA for IGCG technology are tried out to implement it in India.

Experts committees are formed by Central Pollution Control board in consultation with MOEF on court directions. They give directions then and there, (latest being) in 2004, 2005, 2006, for both thermal and Hydel power plant for environmental development.

In the areas of stack emission and effluent discharges, all thermal power plants have been following standard practices, so that pollution under control stages are attained in the sector.

Old plants of private and state utilities are stream lined recently and brought under sustainable technology other wise they are scrapped. New plants are fully incorporated with pollution and environmental control measures.

Damodar Valley Corporation, (DVC) that spread in the present Bihar.

West Bengal, Jhakhand, Chathisgarh states, had taken up measures in 1949 on soil conversation and in afforestation, upland treatment and wasteland treatment on environmental and eco friendly front.

Charter on Corporate Responsibility for Environmental Protection (CREP) by Ministry of Environment and Forests for various industries including thermal power plants are actively pursued by all thermal plants in India including NTPC for achieving sustainable environmental development in the power sector of the country.

Sound Environmental Management System and Practices at the power plant's initial formative stages itself have made it mandatory for strict observance of environmental protection and damage control to obtain clearance from Central Electricity Authority.

Almost all Central public Sector thermal power stations had bagged coveted national and international awards in environmental excellence. State Electricity Board utilities and private utilities are trying to attain this status.

As per above said, initiatives and rules and regulations of the Ministry of Environment and Forests, Government of India, and Ministry of Power, one concludes that there will not be environmental problem of the thermal power utilities in India. Hence the book had not made much of study on environmental aspects.

CONCULSION

The financial performances and other performances of pre and post reforms has shown marked achievements and improved their efficiencies leading to loss reductions of SEB's, higher installed capacities, Power per-capita increases, Higher profits of utilities, environmental achievements, healthy competition among states, more rural electrification of villages, revenue increases and above all the sector had distribution reforms and reduced T&D losses. These performances had lead to a sickness to steadiness condition. This will pay ways for healthiness in future.

5

POLICY IMPLICATIONS FOR THERMAL POWER SECTORS

Thermal power is the main stay in power sector in India. Performance of thermal utilities depend on various efficient operational tools such as quantity of generation units generated in a year, PLF %, capacity additions, fuel consumption, secondary oil consumptions, auxiliary consumptions, forced outages, availability, partial loss, etc.

Thermal utilities performance before reforms and after reforms show improved performance. However, pre reform performances are as low 245.44 billion units per year in 1989-90 as compared to post reforms performances of 617.4 billion units in 2005-06. Pre reforms performances even though have acceleration, their ratings are lower to post reforms. This is due to fuel arrangements, improved technology adopted and better methods of practices implemented.

THERMAL POWER GENERATIONS

Generations of thermal utilities are key success in the power sector. The performance of pre reforms period have a growth of 6 billion units on an average per year. Post reform have generation have an average of 25 billion units per year. The growth acceleration before reforms are marginal due to infant stages and initial technology developmental problems in the sector.

Generation growth of post reforms are due to advancement of technology and better practices, technical advancements, fuel availability, improved PLF, reduction in losses of partial losses and forced outages. Generation growth of 372 billion units in 15 years during post reform period as compared to 240 billion units in 40 years at pre reform period is higher.

There is a better growth in post reforms period on generations especially in thermal as it have a share of 81% in total generation, 500 billion units out of a 617 billion in 2005-06 is a high performance as against 178 billion units out of 245 billion units in 1989-90.

PLANT LOAD FACTORS

An important performance factor. The percentage of PLF if it is higher, then the performance is better. Pre reforms period PLF measures have a slump. In many years PLF are at low levels during 1980's, that contributed less performances. Improvement in PLF started afterwards due to R&M programmes implemented.

Pre reforms PLF growth are very low. During 1976 to 1990 the PLF growth are up and down. Accelerations are negligible with an variation ranging-35 to 7.8%. The slump in PLF are lower performance factors in plants, not adopting better methods and technical methods.

Post reforms PLF acceleration are steady. 1990-91 PLF stood at 56.5% and raise to 73.71% in 2005-06. The acceleration during these period are better. On an average 1.14% are on the increase of PLF from 56.5% in 1990-91 to 73.71% in 2005-06. This is possible due to fuel availability, lower forced outages, higher availability condition, and reduction in preventive maintenance durations, selection of appropriate capacity equipments.

The post reforms PLF growth are gradual and on the accelerated ranges that has improved better performances compared to pre reforms period and paved way for faster growth in generation and profits.

AUXILIARY POWER CONSUMPTION

Auxiliary consumption are the power consumption in % for the power plant generations. The consumption pattern is an economic factor. Reduction in this consumption are a good measure. Higher consumption show lower efficiency.

Pre reform period auxiliary consumption show a high %. This ranged between 9.9 % to 11.1% in the years of1980-1990, Variation are negligible. This is due to non application of reduction measures. Higher consumption pattern. Plant systems and practices, uneconomical usages.

Post reforms period see a variation and reduction. During this period a reduction of 3% are noticed with less fluctuations. The variations are ranged between -1.51% to -10% in the years 19996-97 to 2005-06 except in the year 2002-03 which are 9.5% due to un-foreseen conditions of higher maintenance and consumption. Less auxiliary consumption are due to better maintenance, and selection of better equipments and economic measures taken in the plants. The auxiliary consumption in percentage have decreased over the years.

PREVENTIVE/PLANNED MAINTENANCE

Preventive maintenance percentage reduction is a performance indicator. Among the sectors privates show best performance compared to central and state sector. Private sector show almost one third lower with state and one half lower with central sector indicating their best possible maintenance schedules.

Higher capacity plants such as 500, 250, 210, and 120 MW show less percentage of preventive maintenance as compared to smaller sizes of 110 MW due frequent failures.

Pre reforms higher percentage attracted CEA's attention to take measures for reduction. Post reforms have reduced more than 5% reduction as preventive maintenance and aided more savings in generation and costs with a combined effort of CEA and plants. This is possible by determined scheduling. The PM in percentage have decreased over the years. This is a good indication of reduction in preventive maintenance percentage that lead to performance efficiency

FORCED OUTAGES

Forced outage percentage reduction is a performance indicator. Among the sectors privates show best performance compared to central and state sector. Privates show one third percentage performance lower with state and one half lower with central indicating their best possible forced outage reduction efforts.

Duration wise outages help analyze pitfalls in the outages to further programme in reductions. Mean time analyzes inform capacity wise outages to understand and help planning reduction programmes.

Pre and post reforms forced outages position are indicative of 50 percentages reduction and the improvement made in this outages and subsequent efficiency. Major outages are due to boilers. Improvements in boiler outages and its causes to be further studied and improvements in this needed. However there is a 50% performance efficiency made in forced outages loss percentages in reforms period.

OPERATING AVAILABILITY

Operating availability is a performance indicator. Private sector score more availability of 93.09 as compared to central 86.78 and states 77.52 percentages in 2006, stamping their best efforts. Capacity wise availability show that lower capacities of 10/20, 50 and 60 MW plants have less availability and indicate that higher capacities have good availability performance percentages.

Pre and post reforms performance of this has shown an improvement of 10% on higher side during post reforms. This is a better availability achieved by incorporating and avoiding outages in a systematic way.

Despite its improved performances over last five years in post reforms period, 2005-06 see a slump due to non availability of gas and coal shortages. Gas availability and coal availability to improve operating availability is the important aspects.

SPECIFIC COAL CONSUMPTION

Specific coal consumption is a performance indicator, Per unit consumption of coal quantity. This has been over the years reduced. Pre and post reforms consumption show a reduction of average 0.050 kg per unit. This is possible by various technical and equipment efficiency methods adopted through analyses and systematic approaches. Even a 10 gram of coal saving per Kwh will fetch a substantial saving in cost and fuel. Thermal plants coal consumption efficiency need further reduction to reduce unit cost of power as well as to improve fuel problems.

Coal quality available in regions and its consumption matters much. Mixing of better heat rated coal and good quality coal even if requires imported coal for this purpose as done by Dhanu thermal plant to improve consumption performance.

The coal consumption performance is the base of power generation, selecting a good mix to improve heat rate and coal consumption is the correct solution to improve further the efficiency and increase better consumption performance.

PARTIAL LOSS PERCENTAGE: (DUE TO EQUIPMENT)

Partial losses due to equipments, caused mainly due to boiler and its auxiliary in the high range as compared to turbine and its auxiliary. In the next lowest are the generator. How ever more of loss occurred in miscellaneous mechanical and electrical areas losses. Pre reforms partial losses % are on the high side in equipment causes in the range of 14 to 16%. These are attributed to make of the machine, which have been classified and reported to BHEL and Russian manufacturers to develope and rectify design aspects of major flaws.

Post reforms % show at the range of 13.25 to 8.33% in the reducing level. This has been possible due to scientific study of equipment faults and their history analysis. The Partial Loss in percentage has decreased over the years as the loss prevention by making the equipment in ready and usable condition without repairs and maintenance available for generations.

PARTIAL LOSS PERCENTAGE (DUE TO LOW SYSTEM DEMAND / RESRVE SHUTDOWN)

Less generation by operating units are due to system load variation (Low system Demand). Low load or non utilization of available units are due to their complete shutdown and for want of load (Reserve shutdown) and their main cause are found more in the eastern region at 11.03%. Load prediction and less concerted efforts, planning are the general causes.

Higher loss ranging 5 to 6% in pre reforms as compared to lower 2.1% to 0.92% are made possible due to concentrated load study and balancing acts that are trained with the partial loss reduction among operators and engineers by adopting very accurate predictions.

SECONDARY OIL CONSUMPTION

Fraction of reduction on this Secondary Oil Consumption will save huge cost. Oil seal leakages, flow pattern study have made major saving system. Pre refoms consumption has higher consumption range of 10.6 ml/Kwh has been very much reduced in the post reforms period at 0.63 to 4.36%.

This is possible due to leakage pattern arresting, oil seal technologies development and in-depth consumption study. This is a good effort and savings made resulted in imported commodity, which is more costly and less available item. Development of new substitutes and a good R&D will further reduce the consumption in future.

RENOVATION AND MODERNIZATION (R&M)

This is a step forward on plant operation efficiency after seeing a below normal output of power stations. The effort of renovate and modernize the older plants as a means of augmenting capacity economically improved performance, PLF and more generation.

Various plan periods of 7th, 8th, 9th, 10th Plans Renovation and Modernization programmes added additional generations to the tune of 53550 million units at a very nominal expenses spent on plants. This has given new, replaced and renovated condition that bring more power generation, efficiency, maintenance losses and improved performances.

Despite for the steam lining of all plants the results have encouraged status and condition of plants. The programme has increased life and age of plant and equipments services with an expert team involved in it as national policy.

Post and pre reforms period have a good achievement. However financial arrangements in post reforms period have a slump in this during initial stages are now covered up to have a good financial arrangement guaranteed in the post reforms thrust.

PER-CAPITA ELECTRICITY CONSUMPTION IN INDIA

In the Growth of Indian per-capita capacity consumption, thermal generation have a substantial role in meeting and increasing this consumption. Other power generation sources are very meager contribution with a combined contribution of less than cumulative 19% from five sources such as nuclear, Hydro, cogenerations, renewable, wind. Year wise per-capita consumption increase are possible with the higher level generation of thermal contribution only.

Even lesser developed countries such as middle east nations, and south American nations are having twice the per-capita consumption than India. The consumption pattern increased in domestic, agriculture sectors due to free and cross subsidy of power by governments.

As the public awareness and more use of electricity driven domestic and house hold equipments increased even among villagers, demand and per-capita consumptions have increased very much from 16.5 kwh in 1950 to 616 kwh in 2006. But the major supply source of thermal generation become the backbone of the nation in the off shoot of per-capita increase.

INSTALLED CAPACITY

Installed capacity grown from 107 MW in 1915 to 124287 MW in 2006. Growth of power plant installations since 1915 is marginal. In the first 75 years it is 63500 MW. The growth of power installation since 1991 is very high. It is 58201 MW in just 15 years with a higher average growth. The power sector's major demands of power are met by huge thermal capacity addition only.

Power supply and demand persisted for 13% as peak load demand. Energy loss, commercial losses are hovering around 35 to 40% as national average. Powers produced are unable to meet requirements of growing population demand leaving a huge gap. Power cuts and black out has become order of the day. The only saviors are more installations and bulk installations of thermal plants. To augment this large capacity, and to meet the nations 8% GDP growth, planning and targets needs to be met.

Plants in the size of 4000 MW as ultra mega power projects in higher economies of scales are the latest plan by ministry of power for the XIth plan to meet the target of 212000 MW target achievement in 2012. Installed capacity addition is the only way out to meet power shortages. The supply and demand narrowed only by capacity addition. Any lag in this will affect shortage of power and frequent disturbances.

RURAL ELECTRIFICATION

Growth of rural areas, tapping agricultural and village natural resources are the rural electrification development strategy of India. Reaching out rural area in the power map is a wide gap. Out of 5,87,258 total inhabited villages in the country as per 1991 census, 5,08,863 villages i.e. 87% villages have been declared electrified by March 2002. A balance of 77,897 villages still remain un-electrified since long from, 1897 as the power supply entered Indian scene. India's unelectrified villages are 13% villages numbering 77897 in 2002.

To electrify this 77897 villages, rural electrification programme under Power Finance corporation financial assistance and tapping the thermal power generation are only source of vital linkage. As the rural electrification prograame's new power installations in renewables, cogeneration, small captive generation have yet to develop schemes and projects to meet the requirement not come up due to financial, technical and economical constraints. Augmentation of thermal power generation in the vicinity in a large scale are the only solution to serve this needy village rural electrification programme to complete balance 40,160 villages in 2007.

FINANCIAL PERFORMANCES OF THERMAL POWER UTILITIES

In the wake of poor financial status and financial un-health of state power utilities, the status of financial performance are very a difficult factor to manage. SEB's show a very poor rate of return. Minimum statutory require ment of 16% ROR each year never met. Instead the SEB's Rate of Return on negative side go on increasing in the pre reforms period and in the beginning of reforms period.

1991 to 2000-01 the rate of return increased to -41.5 % accumulating huge losses from 1991 to 2000-01 alone Rs. 65787 crores due to subsidy to farmers, cross subsidy to domestic consumers. At this rate of losses the SEB's nether able to pay the generation power purchase payments or make any new power installation to the growing power need of the nation.

Reforms and commercial losses reduction have played major role containing the huge loss situation and it start a reduction in losses since 2001-02 from a rate of return (RoR) from -37.9 % to -28.13%. Otherwise the SEB's become bankrupt he danger level needs to be sustained to serve the backbone services of the daily citizen needs of the nation, otherwise it may lead to a huge deficit in the national economy. Where as the efficient central and private thermal generation utilities like NTPC, NLC, Reliance Energy have shown a huge profitisation in the range of huge billions of rupees and grow faster. Reform and restructure in SEB's working are the only way to come out of heavy accumulated losses and retrieve for profit, like central and private power generators.

POWER TARIFF

Generation, transmission and distribution and their cost form total cost of power. Tariff are the price fixed by power seller or distributor to end user the customer. Tariff of state SEB's are the main causes of huge losses that dose not meet cost of power supplied to consumer and ultimately ruined the SEB's thermal power generation healthy status to a poor, under developed utility in the long run due to a huge accumulate loss.

The cost of power and tariff/sold price of power in each state for 1975-76 and 2000-01 show wide variation. Cost of power has a variation of 1034.43% to a maximum of 2270%. Tariff/ Sold price variation are 601.9% to a maximum of 1832% only. This has led to a huge gape and leading to more than Rs. 1, 55, 432 crores in between 1991 to 2001-02 in 10 years. The gap are due to 67, 325 crores subsidy provision by SEB's and the balance due to commercial losses and cross subsidy to domestic consumers from a skewed tariff and lower than price or tariff of the actual cost that meet only 68.6% in 2001-02. There is difference between the Real Average Cost of Supply and real tariff of power in SEB's that lead to losses .State and Central regulatory commissions formed in 1998 and set order to rationalize the cost of supply and tariff fixation to each category of consumer. So that tariffs really save the thermal utilities that supply 81% of the nation's power to realize its cost to develop further.

POWER REVENUE

The average cost of supply (ACS) and average revenue realized (ARR) gap is an important measure of financial results per unit of power. The gap between ACS and ARR decreased from 0.38 to 0.20 per unit during 1996-97 to 2004-05 period. The gap as it widen will increase loss. Thermal power plants power generation which is 81% of the total power supplied to customers suffer heavily due to this widening gap. Reforms have initiated steps to contain this gap and narrowing it recently since 2002-03 by tariff hike, collection efficiency and loss reductions.

Hypothesis verifications for average cost of supply and average revenue of SEB's reveal that, the Calculated 't' value is 0.6 the table value is 2.262. The calculated 't' value is lower than the table. Hence the hypothesis gets accepted. There is difference between the Average Cost and Average revenue of SEBs that lead to losses.

The revenue gap narrowing have direct effect on the profits and financial performance of the thermal generation utilities along with distribution utilities to sustain and recover cost of power. Revenues gap decrease in the post reforms period will sustain losses of state thermal utilities to growth and recovery of losses in the years to come.

They should strive to increase revenues by reducing commercial losses and improve collection efficiencies and supply cost reductions. The average cost of supply for 2006-07 are increasing and the revenue realization are touching very near to that, but not above the supply cost. Loss making are reduced much lower compared to previous years due to regulatory commission efforts.

COMMERCIAL LOSSES

Commercial losses at transmission, distribution and pilferage/theft losses in past are on the increase. Commercial losses have raised in all states in 1985-86 Rs 15.2 billions to Rs. 291.3 billions in 2001-02. These losses have an impact on thermal generators on realizing their payment and specially the SEB's and their thermal utilities. Commercial loss reduction in transmission and distribution have an impact of realizing only average 50% of the cost of supply. At this rate of commercial losses which are primarily from theft of power accounts for nearly 20 to 25%. Reduction in this area alone will save Rs. 70 to 75 billions a year. Power distribution un-bundling programme bring new thrust on commercial loss reductions. Power utilities, after reforms resort to unbundling. The un bundling process give importance to identify loss causes, since the distribution utilities are expertise in the task of commercial losses in a circle level and can reduce technical and pilferage, theft of power by deploying high voltage distribution system.

Theft of power is mandated by Electricity Act 2003 to be curbed, will actuate reduction process. In 1990-91 to 2005-06 commercial losses have fluctuations. The reforms and the thrust on losses reduction have brought down since 2002-03 to 2004-05. Huge percentage of commercial losses, reduced to more than 15% on an average. Unbundled utilities on distribution are taking care of this reduction to redeem SEB's and utilitie's financial health in the years to come.

REFORMS AND STATES

Though the reforms initiated in 1991, it took long years to conceptualize the state governments mind set. Only in the year 1996-97 it got serious action after a push from central governments action. In 1998 efforts of regulatory commission setting up, have further initiated reforms momentum.

However reforms stipulation of unbundling SEB's in to different generation, transmission, distribution utilities till 2007 have gone at slow pace. 13 out of 27 states have gone for unbundling. States like Tamil Nadu, Kerala are very slow and require extension of time to unbundled it.

Regulatory commissions are not yet fully geared up to go for tariff rationalization and consumer satisfaction. Circle wise profit center and distribution reforms despite financed by central funds have not yet fully picked up to reform fully.

In order to encourage and bring out motivation, ministry of power have been conducting a competition among states, so that they make efforts in reforming and improving financial viability. This effort is very much drive front running states. However low ranked states are not taking up this competition and challenge in a serious manner. This competition may bound to grow initiatives among states in future, if more attractive incentives are offered to these states.

Various states interrupt reforms in a typical political way to put into action. This in-fact are affecting the major power sector, thermal utilities in regaining their accumulated losses specially by State electricity boards who are caught between political gains of the state governments and employees of these state utilities, who enjoyed benefits for a long period and do not want to be reformed.

The process of reforms are slow and very much detrimental to the growth of the sector and in special reference to thermal utilities who constitute major chunk of Indian power scenario by their generation, realizing power sold to distributors and further develop to meet growing power demand. Reforms had liberalized private sector to set up generation units at ease, so that growing private participation and investment in generation units will narrow demand and power supply. Reforms aims mainly at the financial return and stake holders satisfaction. Reform will pave ways for healthy thermal generation and payment realization.

1. The main stay thermal power utilities in India during power sector reforms have a good growth in capacity addition to more than 1,24,287 MW upto 31.03.2006 marking good growth acceleration.
2. Generations of these sector in overall and in particular on thermal generation have a good growth during reforms, attaining 617 billion units and 500 billion units in 2005-06 respectively to cater national need.
3. Plant performances of thermal area such as plant load factors and other operational efficiencies have improved a lot to set standards in future with a national average PLF of 73.71%, and beyond this.
4. Need for financial performance and overall financial achievements in central and private sector are achieved

with a bench marking high profits. State utilities losses start reducing.

5. In the state sector losses have been considerably reduced. However, still further reduction in losses are expected to follow in the years to come, to set the financial achievements and rate of returns attaining a -12.3 % in 2005-06.
6. Commercial loss pattern start reducing over the years, with reduction in power theft and commercial losses that will help thermal sector very much in their developments.
7. Tariff rationalization leading to power pricing system are established by regulators in states and in centre which will take care of all stake holders such as, generators, loss reduction, consumers for a healthy financial condition for improvement in the power sector.
8. Correct climate of competitive and privatized channel opened up for improving generation and distribution of power sector that start created a beginning.
9. Though the reforms made beginnings in the above areas, it is still in the developmental stages only. Thus much more results and acceleration are bound to come only in the future if it continues in the future with more seriousness pushed by central government.
10. State, central financial assistance, reforming sector have made a head way assistance. It will have budgetary support in future.
11. Eco-friendly approach and attempts are made in environmental and pollution front. This should be made mandatory to all thermal power plants.

IMPLICATIONS

1. Power Sector Reforms have not been fully effective, through out Indian states. They are in the developing stages only, and

in certain states like Tamil Nadu they have not yet been implemented with true spirit. State and central government should make a head way in this.

2. State Electricity Boards are not yet fully implementing unbundling process to impose more effectiveness and efficiency.
3. Tariff of power through regulators should have more influence to rationalize it.
4. Power theft, commercial losses, loss reductions by SEB's have not been seriously addressed and accounted for, by authorities for profit making of the power sector as a whole, this area need much thrust.
5. National back bone, thermal power generation and its fuel resource, such as coal, gas production need a special attention and investment to augment future projections of power and energy security.
6. Strategy should be evolved for compelled installation of more Eco-friendly thermal or other power plants, adhering to international norms.
7. Energy conservation and saving should be a (slogan) mantra to reduce and substitute additional capacity creation.

LOAD SHEDDING IN THE COUNTRY

Recent Tamil Nadu state's notice to cite an example, for not to use grid power by industries in peak hours, according to Indian chamber of commerce and Industry, Coimbatore, reported in March 2008, is an indirect 24% power cut indication, that concerns the industries, (once a surplus power state, in recent years).

Peak shortages were 13.5% during 2006-07 are projected to reach 14-15% and further. States like U.P and Bihar are expected to have a shortage of 30-35%. The causes are lower investments in generation and large AT& C losses. Western region faced higher deficit followed by northern and southern in 2006-07. In order to avert this XI th plan makes an additional capacity plan for 78577

MW and AT&C losses at 15% coupled with energy conservation programmes. UP and Bihar are figured out to have 30-35% and more during May-June. This is the reality.

WIND POWER DEVELOPMENT

India has shown a remarkable progress with 41% and above growth in the last 3 years. Its installed capacity of 1167 MW in 2000 has grown to 6270 MW in 2006 December. Even the world average growth recorded only 8.45%. Tamil Nadu lead the country's wind power capacities with 52% of the nation, followed by Maharastra 20%, Karnataka 12%, Rajastan 7% and Gujarat at 6%. All states offer attractive incentives to make more private players in the field. The Electricity Act 2003 also make concessions without surcharge for captive use. MSEB to be more attractive have come out with free road construction, 50% subsidy for private developers and 50% with out interest if the plant is transferred to MSEB. SERC's of many states opted for intra-state open access regulations. Grid connectivity from wind power although pose problems, with the up coming technologies of flexible AC transmission system, static synchronous compensator and unified power flow controller will make reactive power into system for voltage supports. The wind power development will certainly help to a certain extent (4.5%) to the larger power need of the nation which is slightly better than atomic power source, at current installations.

CAPTIVE POWER PLANTS

The country's estimated captive power capacity are 45,000 MW. Out of this 1 MW and above are estimated as 23,000 MW. In the 11 th plan another 12000 MW may be added. MOP is urging state regulators to form policies for these set-up. 18,000 MW are connected to girid. Various regulatory issues, higher levies, cross-subsidy, Wheeling issues make it non profitable for producers and are discouraged in this category.

The Electricity act removed the need for approval for setting up captive plants from SEB's and paved ways for introducing open access and power trading, but in implementation it lags. This has made un-attractive and un-economical. Rise in oil, crude prices make other alternate energy sources of baggasse, wind, hydro, clean development mechanism, duel fuel waste heat recovery, combined heat system for cost effective generation. If a helping policy are formulated, many new comers will be interested to install in this category to reduce the capacity addition problems of the country.

COAL SOLUTIONS FOR THERMAL POWER

Nations annual coal demand in 2012 are 630-650 Million tons. For power sector alone it consumes 70%. By the year 2020 it may go upto 1267 million metric tons per annum and by 2031-32 it will be 2 billion tons per annum from the current 425 million tons as per coal vision 2025. To augment this investment of $30 billions are required to feed the nation with energy secured. Budget implications in coal for 2007-08 are very low to Rs. 40.63 billions. Deficit can be made available by imports only. Captive coal blocks are the answers that are growing 10.8% rates. Mine developer cum operators as service providers are developing in this field with joint venture set up. Technologically superior foreign companies are gearing up for this set up as an out sourced supplier to power companies. Thus captive coal mining, coal block allocation, blending of imported coal, private participation in the sector can ensure sustainable coal supply for the power capacity addition programmes. As the capacities are primarily added through coal based plants, availability of coal and its management become serious task in the power development, as a primary task.

CONCLUSION

The experience is that the power sector under-goes transformation for economic improvement since 1991, for consumer satisfaction and basic citizen's need. This is a national priority to be taken care in the form of power sector reforms. Improvement in various performances and financial status of the utilities have just started.

As the sector expands larger to meet demands of citizens, various issues of tariff, profitisation, power losses, revenue realization, distribution development, cost effective generation, fuel arrangements, infrastructure development, budgetary and policy development need support from authorities of states, central and united efforts of these two combinations.

In the back drop of current economic reforms and its success, policy makers have made a frantic efforts to contain damage in the sector. The process of continued implementation in line with national power policy will hold key to future development and success in the economy of the 100 billion populations of the nation. State's acute power shortages increases. The development and acceleration are bound to increase with coal, gas, resource crunches with time as we pursue further reforms process in the sector in a serious manner for the entire benefit of the national economy, which is heading for doubling the power capacity in a decade. The task before each one of us is very big, but achievable with dedicated efforts. Power being a national life line, let us power it with new found reformation.

BIBLIOGRAPHY

Adilson de Oliveira (2003), Brazilian Power Industry Reform, The Political Economy of Power Market Reform Conference, Stanford University, Stanford.

Ahluwalia, I. J. and Berry, A. (1996), "Perspectives: Focus on Indo-Canadian Business and Economic Issues", Sponsored by the Canadian International Development Agency, cited in the *Economic Times*, p. 3.

Ahluwalia, M. S. (2001), Report of The Expert Group (on) Settlement of SEB Dues, Ministry of Power, New Delhi.

Ahluwalia, S. S. and Bhatiani G. (2000), Tariff Setting in the Electric Power Sector–Base Paper on Indian Case Study. TERI Conference on Regulation in Infrastructure Services, New Delhi.

Alesina, A. Ardagna, A. Nicoletti, G. and Schiantarelli, F. (2003), "Regulation and Investment", NBER Working Paper Series 9560, National Bureau of Economic Research, March, Cambridge, MA.

Alagh, Y. K. Shah, J. Shah, V. K. (1998), Power Economics in Gujarat, Har Anand Publications, New Delhi.

Amstrong, M. Cowan, S. and Vickers, J. (1994), Regulated Reform: Economic Analysis and British Experience, MIT Press, London.

Amstrong, M. (1997), "Competition in Telecommunications," Oxford, *Review of Economic Policy*, Vol. 13, No. 1, p. 64-82.

Antonette, D'Sa., Murthry, K. V. N. Reddy, A. K. N. (1999), "India's Power Sector Liberalization: An Overview," *Economic and Political Weekly*, Vol. XXXV, No. 40, p. 3012-3018.

Averch, H. and Johnson, L. (1962), "Behavior of the Firm Under Regulatory Constraints", *American Economic Review*, Vol. 52, No. 5, pp. 1052-69.

Anderson, Dennis (1975), Study of Electricity Tariffs in Andhra Pradesh, International Bank for Reconstruction and Development, Washington.

APERC (2002), AP Tariff Order 2002-2003, Andhra Pradesh Electricity Regulatory Commission, Hyderabad.

APGENCO (2003), A Decade of Performance 1992-2002, Andhra Pradesh Generation Company, APGENCO, Generation Particulars, *www.apgenco.com.*

APTRANSCO (2001), Andhra Pradesh Transmission Company, Power Development in Andhra Pradesh 2000-01.

Arentsen, M. J. Fabius, J. W. and Künneke R. W. (2001), "Dutch Business Strategies under Regime Transition", In Midttun, A. (ed.), European Energy Business Strategies, Elsevier, London.

Arulraj, A. Rajasekaran, D. (2006), Towards BIMSTEC-Japan Comprehensive Economic Cooperation: Vision and Tasks Ahead (Energy), Organised by CSIRD, KolKata & SP Foundation, Tokyo International Conference, Kolkata.

Annual Report (2006), Ministry of Power, Annual Report, Government of India, New Delhi.

Bacon, R. W. and Besant-Jones, J. E. (2001), "Global Electric Power Reform, Privatization and Liberalization of The Electric Power Industry in Developing Countries," Annual Review of Energy and The Environment, pp. 331-359.

Baijal, P. (1999), "Restriction Power Sector in India," *Economic and Political Weekly,* Vol. XXXIV, No. 39, pp. 2795-2803.

Beato and Laffon (2002), World Bank-1997, The World Bank's Role in the Electric Power Sector, The World Bank, Washington, DC.

Bharadwaj, A. (2002), Gasification and Combustion Technologies of Agro-Residues and Their Application To Rural Electric Power Systems in India. Engineering & Public Policy, Mechanical Engineering, Carnegie Mellon University, Pittsburgh.

Bharadwaj, A., Tongia, R. (2003), "Distributed Power Generation: Rural India–A Case Study," IEEE Transactions on Power Systems preparing for Publication Submission.

Bialek, J. W. (2004), "Recent Blackouts in US and Continental Europe: Is Liberalisation to Blame?" CMI Electricity Project Working Paper No. 34, Department of Applied Economics, University of Cambridge.

Bidwai, P .(2002), The other Enrons, *Frontline,* Vol. 19, No. 4, p. 100-106, Chennai.

Blue Print for Power Sector Development (2001), Ministry of Power, Government of India, New Delhi.

Borenstein, S. (2001), "The Trouble with Electricity Markets (and some solutions)", Program on Workable Energy Regulation (POWER), Working Paper, PWP-081, January, University of California Energy Institute, Berkeley.

Bosi, M. (2000), An Initial View on Methodology for Emission Baselines: Electricity Generation case study, IEA Information Paper, Energy and Environment Division, IEA, Paris.

Bower, J. (2002), "Seeking the Single European Electricity Market: Evidence form an Empirical Analysis of Wholesale Market Prices," Working Paper EL 01, Oxford Institute for Energy Studies, London.

Boisseleau, F. (2004), "The Role of Power Exchanges for the Creation of a Single European Electricity Market: Market Design and Market Regulation," PhD Thesis, University of Paris IX Dauphine, Delft University Press.

Bortolotti, B. and Siniscalco, D. (2004), The Challenges of Privatization: An International Analysis, Oxford University Press, Oxford.

BP (2002), British Petroleum Statistical Review of World Energy: Natural Gas, British Petroleum.

Brown, R. E. Marshal, M. W. (2001), "The Cost of Reliability," Transmission & Distribution World.

Brunekreeft, G. (2002), "Regulation and Third-Party Discrimination in the German Electricity Supply Industry," *European Journal of Law and Economics* Vol. 13, p. 203-220.

Burdon, I. (1998), "Options for Mid-Merit Power Generation in The UK Electricity Market", *Power Engineering Journal*, London.

CEA 15th and 16th Electric Power Survey of India, Central Electricity Authority, New Delhi.

CEA (1997), Fourth National Power Plan, Central Electricity Authority, New Delhi.

CEA (2000), Fifth National Power Plan, Central Electricity Authority, New Delhi.

CERC (2001), Draft Competitive Bidding Regulations, by Price Water House Coopers for Central Electricity Regulatory Commission, New Delhi.

Cerra, V. Saxena, S. C. (2002), "What caused the 1991 Currency Crisis in Indian?" IMF Staff papers Vol. 3 pp.. 49.

Chang (1997), Financial Integration with and without International Policy Coordination , *International Economic Review*, Vol. 38, pp. 547-564.

Chidambaram, R. Ganguly, C. (1996), "Plutonium and Thorium in the Indian Nuclear Programme," *Current Science,* Vol. 70 (1) p. 23,35.

Choukroun, S. (2001), Enron in Maharashtra: Power Sector Development and National Identity in Modern India, University of Pennsylvania, Philadelphia.

Churchill, A. A. Saunders, R. J. (1989), "Financing of the Energy Sector in Developing Countries", 14th Congress of the World Energy Conference, Montreal.

Clinch, J. P. Healy, J. D. (2000), Cost-Benefit Analysis of Domestic Energy Efficiency, *Energy Policy,* Vol. 29, No. 2, p. 113-124.

CRISIL and ICRA, (2006), "Rating of The State Power Sector", for Ministry of Power Government of India, New Delhi.

Domah, P. D. and Pollitt M. (2001), "The Restructuring and Privatisation of the Regional Electricity Companies in England and Wales: A Social Cost Benefit Analysis," *Fiscal Studies,* Vol. 22(1), pp. 107-146.

Drillisch, J. and Riechmann, C. (1998), "Liberalisation of the Electricity Supply Industry: Evaluation of Reform Policies" EWI Working Paper 98/5, Cologne/Tokyo.

D'Sa, A. (2002), Power Sector Reform in India–An Overview, International Energy Initiative, Bangalore.

Deepak Parekh Expert Committee on State Specific Reforms (2002), Structuring of APDRP, Reform Framework and Principles of Financial Restructuring of SEBs, Ministry of Power, New Delhi.

Dhall, A. R. Mirajkar, *et al.* (2007), Institutional Restructuring: What Role can CEA Play, Indian Institute of Management, Ahmedabad.

Dixit, S. Sant, G. (1997), "How Reliable are Agricultural Power Use Data?" *Economic and Political Weekly,* Vol. XXXII, No. 37, pp. 2452-2459.

Dixit, S. Sant, G. *et al.* (1998), "Regulation in the WB-Orissa Model: Curve Worse Than Disease," *Economic and Political Weekly,* Vol. XXXIII, No. 17, pp. 944-949.

DTI (2001), UK Energy Sector Indicators 2001, Department of Trade and Industry, London.

Dubash, N. K. and Rajan, S. C. (2001), The Politics of Power Sector Reform in India. World Resources Institute, Washington, DC.

Dubash, N. K. (2002), Power Politics: Equity and Environment in Electricity Reform, World Resources Institute, Washington, DC.

Dussan, M. (1996), "Electric Power Sector Reform in Latin America and the Caribbean," Working Papers Series IFM-104.

Dunkerly Joy, C. (1996), China's Energy Strategy: Economic Structure, Technological Choices, and Energy Consumption, *Energy Journal*, Vol. 17, Issue 3, pp. 109-111.

EA (2002), Electricity Industry Review 6, The Electricity Association, London.

EA (2002), Who Owns Whom in the UK Electricity Industry, The Electricity Association, London.

Enron Action Group (2001), The Enron Saga, Himalaya Publication, Mumbai.

European Commission (2004), "Third Benchmarking Report on the Implementation of the Internal Electricity and Gas Market." Working Paper, DG TREN, Commission of the European Communities, March, Brussels.

Fischer, R. and Serra, P. (2000), Regulating the Electricity Sector in Latin America, Centre for Economic Applications, University de Chile, Chile.

Foster, C. D. (1992), Privatisation, Public Ownership, and the Regulation of Natural Monopoly, Blackwell, Oxford.

Garg, A. Shukla, P. R. (2002), Emissions Inventory of India, TATA-McGraw Hill, New Delhi, India, pp. 278.

Genoud, C. and Finger, M. (2002), "Regulatory Convergence? The Example of the European electricity sector," Working Paper de l'IDHEAP 8/2002, Institute de Hautes Etudes en Administration Publique, Lausanne.

Girish Sant, Shantanu Dixit (May 1996), "Agricultural Pumping Efficiency in India: The Role of Standards," Energy for Sustainable Development, The *Journal of International Energy Initiative*, Volume II, pp. 1-12.

Gilardi, F. (2003), "Delegation to Independent Regulatory Agencies in Western Europe: A Cross-Sectional Comparison," University of Lausanne, Edinburgh.

Green, R. J. and Newbery, D. (1992), "Competition in the British Electricity Spot Market," *Journal of Political Economy*, No. 100(5), pp. 929-53.

Glover, J. D. Sarma, M. S. (2002), Power System Analysis and Design, Wadsworth, Thomson Learning, Pacific Grove, CA.

Government of India (2002), Census of India, Registrar General & Census Commissioner, New Delhi.

Gupta, A. (2002), A Requiem for Reform, *Business Today*, Vol. 11, pp. 60-64.

Gunn Calum (1997), Energy Efficiency Vs Economic Efficiency?, *Energy Policy*, Vol. 25, Issue No. 2, pp. 241-254.

Heckman James and Richard Robb (1985), "Alternative Methods of Evaluating the Impact of Interventions." In James Heckman and Burton Singer, eds.,

Longitudinal Analysis of Labmy Market Data, Cambridge University Press, New York, pp. 156-245.

IADB (2001), Sustainability of the Electricity Sector Reforms in Latin America, *www.iadb.org*, IADB Research Department, Chile, Santiago.

ICRA/CRISIL (2003), "Rating of The State Power Sector", for Ministry of Power Government of India, New Delhi.

IDFC (2000), Orissa Power Sector Reform: A Brief Overview of the Process, Infrastructure Development and Finance Company, Ltd., New Delhi.

IEA (1998), Regional trends in Energy efficient Coal-fired power Generation Technologies, Coal Industry Advisory Board, IEA, France, pp.190.

IEA (2003), World Energy Investment Outlook: 2003 Insights, International Energy Agency, Paris.

Infraline (2002), Infraline Detailed Power Sector, *Newsletter*, Vol. XX, IT Task Force Report for Power Sector, IT Task Force, Government of India, New Delhi.

Jamasb, T. (2002), "Reform and Regulation of the Electricity Sectors in Developing Countries" DAE Working Paper 0226, CMI EP 08, Department of Applied Economics, University of Cambridge, Cambridge.

Jamasb, T. Mota, R. Newbery, D. and Pollitt, M. (2004), "Electricity Sector Reform in Developing Countries: A Survey of Empirical Evidence on Determinants and Performance." Cambridge Working Papers in Economics 0439(CMI EP47), Department of Applied Economics, University of Cambridge.

Jamasb Tooraj, Littlechild, Stephen, C. (2004), Reform and Regulation of the Electricity Sectors in Developing Countries, Cambridge Working Papers in Economics, 1810/323, Cambridge.

Jamasb, T. (2002), "Reform and Regulation of the Electricity Sectors in Developing Countries." DAE Working Paper 0226 (CMI EP 08), Department of Applied Economics, University of Cambridge.

Jamasb, T. M. Pollitt (2001), "Benchmarking and Regulation: International Electricity Experience." *Utilities Policy*, Vol. 9(3), pp. 107-130.

Johansson, T. B. and Turkenburg, W. (2004), "Policies for Renewable Energy in the European Union and its Member States: An Overview", *Energy for Sustainable Development*, Vol. VIII(1): pp. 5-24.

Joskow, P. L. (1998), "Electricity Sectors in Transition", *Energy Journal* No. 19 pp. 25-62.

Joskow, P. L. (1998), "Regulatory Priorities for Reforming Infrastructure Sectors in Developing Countries," The 10th Annual Bank Conference on Development Economics, World Bank, Washington, D.C.

Joskow, P. L. (2002), "Electricity Sector Restructuring and Competition: A Transaction-Cost Perspective," The Economics of Contracts, Cambridge University Press, Cambridge.

Kandula Subramanian (2004), Power Sector Review, Indian Power Sector, Center for Development Studies, Trivandrum.

Kannan, K. P. Pillai, N. V. (2002), The Aetiology of the Inefficiency Syndrome in the Indian Power Sector, Center for Development Studies, Trivandrum.

Kanungo Committee, Government of Orissa (2001), Report of the Committee on Power Sector Reform of Orissa, Ministry of Power Government of India.

Khanna, M. Zilberman, D. (1999), Barriers to Energy Efficiency in Electricity Generation in India, *The Energy Journal*, Vol. 20 No. 1, pp. 25-41.

Khanna, M. Zilberman, D. (2001), Adoption of Energy Efficient Technologies and Carbon Abatement: The Electricity Generation Eector in India, *Energy Economics*, No. 23 (2001), pp. 637-658.

Kulkarni, V. S. (2004), Power Sector Policies and Perspectives in the Post Reform Period, Abhijeet Publication, New Delhi, pp. 319-331.

Laurin, Claude and Yves Bozec (2000), "Privatization and Productivity Improvement: The Case of Canadian National (CN)," working paper, Ecoles de HEC, Montreal.

La Porta, Rafael and Florencio L-de-Silanes (1999), "Benefits of Privatization—Evidence From Mexico, *Q. J. Econ.*, Vol. 114:4, pp. 1193-1242.

Mahalingam, S. (2001), A Worrisome Tariff Order, *Frontline*, Vol. 18, No.6 pp. 18-22, Chennai.

Mahalingam, S. (1997), Unbundling Trouble, *Frontline*, Vol. 14, No. 8, pp. 84- 89, Chennai.

Mahalingam, S. (1998), Power to the Regulator, *Frontline*, Vol. 15, No. 10, pp. 110-115, Chennai.

Mahalingam, S. (2002), Markets through The Backdoor, *Frontline*, Vol. 19, No. 4, pp. 41-48, Chennai.

Mathur, R. Chand, S. Tezuka, S. (2003), Optimal Use of Coal for Power Generation in India, *Energy Policy*, Vol. 31, pp. 319-331.

Mehta, A. (2000), Power Play: A Study of the Enron Project, Orient Longman, Mumbai.

Megginson William and Jeffry Netter (2001), "From State to Market: A Survey of Empirical Studies on Privatization," *Journal of Economic Literature* No. 39, pp. 321-389.

Millan, J. Lora. E. Micco, A. (2001), Sustainability of the Electricity Sector Reforms in Latin America, IADB Research Department, Chile, Santiago.

Ministry of Coal (2002), Coal Profile, Ministry of Coal, Government of India, New Delhi.

Ministry of Finance (2002), Economic Survey, Ministry of Finance (Economic Division), Government of India, New Delhi.

Ministry of Power (2002), Annual Report, Ministry of Power, Govt. of India, New Delhi.

Ministry of Power (2002), Distribution Policy Committee Report, Ministry of Power, New Delhi.

Ministry of Power (2002), Distribution Reforms Report, Ministry of Power, New Delhi.

Ministry of Power-Media Source (2005), Government of India, *www.powermin.nic.in*.

Ministry of Power (2003), Annual Report, Ministry of Power, Government of India, New Delhi.

Morris, S. Ed. (2002), Indian Infrastructure Report, Governance Issues for Commercialization, 3i Network, Ahmedabad.

Newbery, D. and Pollitt M. (1997), "Restructuring and Privatisation of the CEGB-Was It Worth?", *Journal of Industrial Economics,* Vol. 45(3), pp. 269-304.

Newbery, D. M. (1999), Privatization, Restructuring, and Regulation of Network Industries, MIT Press, London.

Newbery, D. M. (2002), "Improved Revenue Collection", Issues and Options for Restructuring Electricity Supply Industries, European Economic Review Elsevier.

OERC (2003), Orissa Tariff Order, Orissa Electricity Regulatory Commission, Bhubhaneshwar.

OFGEM (2001), Annual Report, Office of Gas and Electricity Markets, London.

OFGEM (2002), Review of Domestic Gas and Electricity Competition and Supply Price Regulation: Conclusions and Final Proposals, Office of Gas and Electricity Markets, London.

OFGEM (2002), Mergers in the Electricity Distribution Sector: Policy Statement, Office of Gas and Electricity Markets, London.

Padmanaban, S. Totino, J. (2001), Energy Efficiency in Indian Agriculture. USAID–India, New Delhi.

Okten, Cagla and Peren Arin (2003), "How Does Privatization Affect Efficiency, Productivity and Technological Choice? Evidence from Turkey," Unpublished manuscript, Louisiana State University.

Parikh, K. S. (1996), The Enron Story and Its Lessons, Indira Gandhi Institue of Development Research, Mumbai.

Paulina Beato, Jean-Jacques Laffont (2002), Competition in Public Utilities in Developing Countries, Sustainable Developing Department-Technical Papers Series , Inter-American Developing Bank, Washington, D.C.

Planning Commission (2002), Annual Report, on the Working of State Electricity Boards & Electricity Departments, Planning Commission, Govt. of India, New Delhi.

Power Grid (2002), Indian Electricity Grid Code, Power Grid Corp, New Delhi.

Power Line (2006), Monthly Magazine on Power, New Delhi, March to November Issues, New Delhi.

Pollitt, M. (1999), "Issues in Electricity Market Integration and Liberalization, in European Market for Electricity?" Centre for Economic Policy Research, London.

Pollitt, M.G. (2004). "Electricity Reform in Chile: Lessons for Developing Countries," CMI Electricity Project Working Paper, Department of Applied Economics, University of Cambridge.

Prayas (2001), India Power Sector Reforms Update, Issue No. 1, Prayas, Pune.

Prayas (2001), India Power Sector Reforms Update, Issue No. 3, Prayas, Pune.

Rahul Tongia (2003), Indian Power Sector Reform Studies, Stanford-CMU Stanford.

Rajan, A. T. (2000), "Power sector reform in Orissa: An Ex-pert Analysis of the Causal Factors," *Energy Policy,* Vol. (10), pp. 28.

Razavi, H. (1996), Financing Energy Projects in Emerging Economies, Penn Well Publishing, Tulsa.

Reddy, M. T. (2000), Development in the Power Sector in Andhra Pradesh Prayas-Focus on Global South–Event on Power Sector Reforms, Mumbai.

Salies, E. and Waddams Price C. (2004), "Charges, Costs, and Market Power: The Deregulated UK Electricity Retail Market." *The Energy Journal,* No. 25(3), pp. 19-37.

Sankar, T. L. (2002), "Towards a People's Plan for Power Sector Reform", *Economic and Political Weekly,* XXXVII, No. 28, pp. 4143-4151.

Sankar, T. L. and Ramachandra, U. (2000), "Regulation of the Indian Power Sector", *ASCI Journal of Management*, Vol. 2, pp. 2028-2034.

Sant, G. S. Dixit, *et al.* (1995), The Enron Controversy, Techno-Economic Analysis and Policy Implications, Prayas, Pune.

Satyanarayan Chary, T. (2001), June, Power Sector Reforms in Andhra Pradesh, *Indian Journal of Public Enterprises*, Vol. 16(30), pp. 68-78.

Shukla, P. R. Debashish Biswas, Tirthankar Nag, Amee Yajnik, Thomas Heller and David G. Victor (2004), Working Paper 20, Program on Energy and Sustainable Development, Center for Environmental Science and Policy, Stanford University, Stanford, CA.

Shukla, P. R., Sharma, S. K., Venkata Ramana, P. (2002), Climate Change and India, Tata McGraw- Hill, New Delhi, India, pp. 317.

Sharma, V. K. (2004) Electricity India Vision - 2007, Galgotia Publications, New Delhi.

Singh, J. (1991), Plant Size and Technical Efficiency in the Indian Thermal Power Industry, *Indian Economic Review*, Vol. XXVI, No. 2, pp. 239-252.

Sridhar, V. (2000), Brutal Crackdown, *Frontline*, Vol. 17, No.10 pp. 38-43, Chennai.

Srinivasan, R. (2002), VSNL Imbroglio: Crossed Wires, *The Hindu* , Businessline, Chennai.

Stern David, I. (2000), "A Multivariate Cointegration Analysis of the Role of Energy in the US Macroeconomy," *Energy Economics, Elsevier*, Vol. 22(2), pp. 267-283.

Stiglitz, J. E. (2002), Globalization and its Discontents, Norton, New York.

Standard and Poor's (2003), Compustat Global Database: MacGraw-Hill Companies, London.

Surrey, J. (1996), From Public to Private Ownership, Objectives and Rationale, The British Electricity Experiment, Earthscan, London.

Sury, M. M. (2001), India: A Decade of Economic Reforms 1991-2001, New Century Publications, New Delhi, India, pp. 253.

Sweeney, J. (2002), The California Electricity Crisis, Hoover Institution, Stanford.

Thatcher, M. (2002), "Regulation After Delegation: Independent Regulatory Agencies in Europe," *Journal of European Public Policy*, 9(6), pp.954-972.

The Ministry of Power (2005), Annual Report, Ministry of Power, Government of India, New Delhi.

The Political Economy of Power Market Reform Conference Report, (2003), Stanford University, Stanford.

The World Bank (1995), Learning from Narmada, The World Bank (Operations Evaluation Department), Washington, DC.

Thomas C. Heller, David G. Victor (2003), The Political Economy of Power Sector Reform: The Experiences of Five Major Developing Countries, Published by Cambridge University Press, Cambridge.

Tooraj Jamasb (2002), Between the State and Market: Electricity Sector Reform in Developing Countries, Faculty of Economics, University of Cambridge, Cambridge.

Tongia, R. (1998), Demand and Supply of Power in India: An Analysis of the Electric Power Grid (working paper) Electrical and Computer Engineering, Engineering and Public Policy. Car negie Mellon University, Pittsburgh.

Tongia, R. and Banerjee, R. (1998), "Price of Power in India", *Energy Policy,* Vol. 26(7), pp. 557-575.

Tongia, R. and Arunachalam, V. S. (1998), "India's Nuclear Breeders: Technology, Viability, and Options," *Current Science,* Vol. 75(6): pp. 549-558.

Tongia, R. and Arunachalam, V. S. (1999), "Natural Gas Imports for South Asia: Pipelines or Pipedreams?", *Economic and Political Weekly,* XXXIV No.18, pp. 1872-1879.

TND (2002), Utilities Policy, Between the State and Market: Electricity Sector Reform in Developing Countries, Cambridge University, *Business Line,* Vol. 14, Issue 1, pp. 14-30.

UCTE (1999), Power and Energy Balances of the UCTE–Retrospect of the Year, Union for the Co-ordination of Transmission of Electricity, Brussels.

UCTE (2000), Power and Energy Balances of the UCTE–Retrospect of the Year, Union for the Co-ordination of Transmission of Electricity, Brussels.

UCTE (2003), December, UCTE System Adequacy Forecast, Union for the Co- ordination of Transmission of Electricity, Brussels.

Upadhyay Anil, K. (2000), Power Sector Reforms, Indian Experience and Global Trends, *Economic and Political Weekly,* XXXV, No. 12, pp. 1023-1028.

Vasconcelos, J. (2004), Services of General Interest and Regulation in the EU

Energy Market, Council of European Energy Regulators (CEER), Presentation at XVI CEEP Congress, 2004, Leipzig.

Varone, F. Aebischer, B. (2001), Energy Efficiency, The Challenges of Policy Design, *Energy Policy,* Vol. 29, pp. 615-629.

Verhoef, E. T. Nijkamp, P. (1999), "Second-Best Energy Policies for Heterogeneous Firms," *Energy Economics,* Vol . 21, pp. 111-134.

Vickers, J. and Yarrow G. (1988), Privatization: An Economic Analysis. MIT Press, Cambridge.

Weir, C. (1999), "Regulation and the Development of Competition in the U. K., Gas Supply Industry", *Review of Industrial Organization*, Vol. 15, pp. 135-147.

Wolak, F. A. (2001), "Market Design and Price Behaviour in Restructured Electricity Markets: An International Comparison," Department of Economics, Stanford University.

Wolsink, M. (2000), "Wind Power and the NIMBY-Myth: Institutional Capacity and the Limited Significance of Public Support," *Renewable Energy*, Vol. 21(1), pp. 49-64.

World Bank (1993), The World Bank's Role in the Electric Power Sector, The World Bank, Washington, DC.

World Bank (1995), Learning from Narmada, The World Bank (Operations Evaluation Department), Washington, DC.

World Bank (1997), The World Bank's Role in the Electric Power Sector, The World Bank, Washington, DC.

World Bank (2001), India–Power Supply to Agriculture, Energy Sector Unit, South Asia Regional Office, Washington, DC.

World Gas Intelligence (2002), Petronet LNG's Ras Gas Deal, Himalayan Publication, Mumbai.

Zhang, C. May, Michael, M. Heller, Thomas, C. (2001), Impact on Global Warming of Development and Structural Changes in the Electricity Sector of Guangdong Province, China, *Energy Policy*, Vol. No. 29(2001), pp179-203.

ABBREVIATIONS

ABT	–	Availability Based Tariff
ACS	–	Average Cost of Supply
AP	–	Annual Plan (or the state of Andhra Pradesh)
ARR	–	Annual Revenue Requirement
BHEL	–	Bharat Heavy Electricals Limited
BJP	–	Bharatiya Janata Party(Ruling party in the present India Government–in a coalition)
BU	–	Billion Units
CEA	–	Central Electricity Authority
CCGT	–	Combined cycle gas turbine
CERC	–	Central Electricity Regulatory Commission
CM	–	Chief Minister
Crore (or cr.)	–	10,000,000 (4.8 crore rupees = 1 million US$ today)
DFID	–	Department for International Development (UK)
DPC	–	Dabhol Power Company
DSM	–	Demand Side Management
DVB	–	Delhi Vidyut Board
EPRI	–	Electric Power Research Institute (US)

ERC	–	Electricity Regulatory Commission (Includes variants like CERC, SERC, OERC, etc.)
GOI	–	Govt. of India
IAS	–	Indian Administrative Services
IPP	–	Independent Power Producer
kWh	–	Kilowatt-hour
MIS	–	Management of information Systems
MoU	–	Memorandum of Understanding
MU	–	Million Units
MW	–	Megawatt
NTPC	–	National Thermal Power Corporation.
O&M	–	Operation and Maintenance
PLF	–	Plant Load Factor
PPA	–	Power Purchase Agreement
ps/kWh	–	Paise per kilowatt-hour (100 paise = 1 rupee)
PSU	–	Public Sector Unit
RE	–	Revised Estimate
RLDC	–	Regional Load Dispatch Center
RoE	–	Return on Equity
RoR	–	Rate of Return
Rs.	–	Rupees
R&M	–	Renovation and Modernization
SEB	–	State Electricity Board
SERC	–	State Electricity Regulatory Commission
SHR	–	Station Heat Rate
T&D	–	Transmission and distribution
WB	–	World Bank
WB-SAR	–	World Bank–Staff Appraisal Report

